Momet

Algebra 1 Workbook

1,400+ Questions

Key Algebra Concepts and Practice

Detailed Answer Explanations

Copyright © 2025 by Mometrix Media LLC

All rights reserved. This product, or parts thereof, may not be reproduced, stored in a retrieval system, or transmitted in any form or by any means—electronic, mechanical, photocopy, recording, scanning, or other—except for brief quotations in critical reviews or articles, without the prior written permission of the publisher.

Written and edited by Mometrix Test Prep

Printed in the United States of America

This paper meets the requirements of ANSI/NISO Z39.48-1992 (Permanence of Paper).

Mometrix offers volume discount pricing to institutions. For more information or a price quote, please contact our sales department at sales@mometrix.com or 888-248-1219.

Mometrix Media LLC is not affiliated with or endorsed by any official testing organization. All organizational and test names are trademarks of their respective owners.

Paperback
ISBN 13: 978-1-5167-2752-0
ISBN 10: 1-5167-2752-5

TABLE OF CONTENTS

INTRODUCTION (READ THIS FIRST!) _____ 1
   ONLINE QUIZZES _____ 1
   QUIZ ANSWERS _____ 1
   OVERCOMING TEST ANXIETY _____ 1

CHAPTER 1: SOLVING EQUATIONS _____ 2
   1.1 SOLVING ONE-STEP EQUATIONS _____ 2
   1.2 SOLVING TWO-STEP EQUATIONS _____ 5
   1.3 SOLVING EQUATIONS USING THE DISTRIBUTIVE PROPERTY _____ 8
   1.4 SOLVING EQUATIONS WITH VARIABLES ON BOTH SIDES _____ 11
   1.5 SIMPLIFYING ALGEBRAIC EXPRESSIONS _____ 14
   1.6 SOLVING EQUATIONS BY COMBINING LIKE TERMS _____ 17
   1.7 SOLVING MULTI-STEP EQUATIONS _____ 20
   1.8 SOLVING EQUATIONS INVOLVING ABSOLUTE VALUE _____ 23
   1.9 SOLVING FORMULAS FOR SPECIFIC VARIABLES _____ 26
   1.10 WRITING AND SOLVING EQUATIONS INVOLVING DIRECT VARIATION _____ 30

CHAPTER 2: FUNCTIONS AND SEQUENCES _____ 33
   2.1 EVALUATING FUNCTIONS _____ 33
   2.2 DOMAIN AND RANGE OF LINEAR FUNCTIONS _____ 36
   2.3 MANIPULATING LINEAR FUNCTIONS _____ 42
   2.4 WRITING FUNCTIONS FOR REAL-WORLD PROBLEMS _____ 46
   2.5 ARITHMETIC SEQUENCES _____ 49
   2.6 GEOMETRIC SEQUENCES _____ 52

CHAPTER 3: LINEAR EQUATIONS _____ 55
   3.1 SLOPE AND RATE OF CHANGE _____ 55
   3.2 GRAPHING LINEAR EQUATIONS _____ 61
   3.3 FINDING THE SLOPE-INTERCEPT FORM EQUATION OF A LINE _____ 72
   3.4 FINDING THE POINT-SLOPE FORM EQUATION OF A LINE _____ 76
   3.5 FINDING THE STANDARD FORM EQUATION OF A LINE _____ 79
   3.6 CONVERTING BETWEEN STANDARD FORM AND SLOPE-INTERCEPT FORM _____ 82
   3.7 WRITING LINEAR EQUATIONS _____ 85
   3.8 PARALLEL AND PERPENDICULAR LINES AND THE AXES _____ 91
   3.9 EQUATIONS OF PARALLEL LINES _____ 94
   3.10 EQUATIONS OF PERPENDICULAR LINES _____ 97
   3.11 CALCULATING CORRELATION COEFFICIENT _____ 100
   3.12 ASSOCIATION, CORRELATION, AND CAUSATION _____ 105

CHAPTER 4: INEQUALITIES _____ 108
   4.1 INEQUALITY BASICS _____ 108
   4.2 SOLVING ONE-STEP INEQUALITIES _____ 111
   4.3 SOLVING MULTI-STEP INEQUALITIES _____ 115
   4.4 SOLVING COMPOUND INEQUALITIES _____ 119
   4.5 SOLVING INEQUALITIES INVOLVING ABSOLUTE VALUES _____ 123
   4.6 WRITING LINEAR INEQUALITIES _____ 127
   4.7 GRAPHING LINEAR INEQUALITIES _____ 134

CHAPTER 5: SYSTEMS OF EQUATIONS — 149
- 5.1 The Graphing Method — 149
- 5.2 The Substitution Method — 165
- 5.3 The Elimination Method — 168
- 5.4 Writing Systems of Linear Equations — 171
- 5.5 Graphing Solutions to Systems of Linear Inequalities — 181

CHAPTER 6: POLYNOMIALS — 196
- 6.1 Adding and Subtracting Polynomials — 196
- 6.2 Multiplying Monomials — 199
- 6.3 Dividing Monomials — 202
- 6.4 Multiplying Polynomials by Monomials — 205
- 6.5 Multiplying Binomials — 208
- 6.6 Dividing Polynomials by Monomials — 211
- 6.7 Dividing Polynomials by Binomials — 214
- 6.8 Factoring Using Common Monomial Factors — 217
- 6.9 Factoring Trinomials of the Form — 220
- 6.10 Factoring the Difference of Two Squares — 223
- 6.11 AC Method — 226
- 6.12 Multistep Factoring — 229

CHAPTER 7: ALGEBRAIC FRACTIONS — 232
- 7.1 Simplifying Algebraic Fractions — 232
- 7.2 Solving Proportions Containing Algebraic Fractions — 236
- 7.3 Multiplying Algebraic Fractions — 240
- 7.4 Dividing Algebraic Fractions — 244
- 7.5 Adding and Subtracting Algebraic Fractions — 248
- 7.6 Adding and Subtracting Algebraic Fractions with Binomial Denominators — 252
- 7.7 Solving Equations Involving Algebraic Fractions — 256

CHAPTER 8: QUADRATIC FUNCTIONS — 259
- 8.1 Domain and Range of Quadratic Functions — 259
- 8.2 Writing Equations Given Vertex and a Point — 268
- 8.3 Factoring Quadratic Equations — 271
- 8.4 Writing Quadratic Functions Given Solutions and Graphs — 274
- 8.5 Solving Quadratic Equations Using Square Roots — 289
- 8.6 Completing the Square — 292
- 8.7 Converting Between Standard and Vertex Forms — 295
- 8.8 The Quadratic Formula — 298
- 8.9 The Discriminant — 301
- 8.10 Graphing Quadratic Functions — 304
- 8.11 Manipulating Quadratic Functions — 319

CHAPTER 9: EXPONENTIAL FUNCTIONS — 323
- 9.1 Domain and Range of Exponential Functions — 323
- 9.2 Writing Exponential Functions — 326
- 9.3 Graphing Exponential Functions and Determining Key Features — 329

CHAPTER 10: RADICAL EXPRESSIONS — 343
- 10.1 Simplifying Radical Expressions — 343
- 10.2 Solving Equations Involving Radicals — 346
- 10.3 Adding and Subtracting Radical Expressions — 349

10.4 MULTIPLYING RADICAL EXPRESSIONS _____ 352
10.5 DIVIDING RADICAL EXPRESSIONS _____ 355
10.6 RATIONALIZING THE DENOMINATOR _____ 358
10.7 SIMPLIFYING RADICAL EXPRESSIONS WITH BINOMIAL DENOMINATORS _____ 362

PRACTICE TEST #1 _____ 366
ANSWER KEY AND EXPLANATIONS _____ 384
PRACTICE TESTS #2 AND #3 _____ 414
QUIZ ANSWER KEY _____ 415
HOW TO OVERCOME TEST ANXIETY _____ 420
ADDITIONAL BONUS MATERIAL _____ 426

Introduction (Read This First!)

Online Quizzes

This guide is broken down by chapter and by topic. Each topic consists of a set of quiz questions to work out. These quizzes can be taken on paper or on almost any internet-connected device via our online **interactive testing interface**, which can be accessed by visiting the bonus page for this product at **mometrix.com/bonus948/algebra1wb** (or via the QR code) and clicking on **Lesson Quizzes**.

Quiz Answers

An answer key for the quizzes is located at the end of this book, but in an effort to keep the book to a manageable length, we've opted to provide the complete worked out explanations for the quiz questions electronically only. So if you prefer to take the quizzes on paper, you'll want to visit the link below to see the **worked out answer explanations**. Visit the bonus page for this product at **mometrix.com/bonus948/algebra1wb** (or via the QR code) and click on **Quiz Answers** to access the PDF showing how to work out the answer to each question. The PDF has a clickable table of contents for easy navigation to the right quiz.

Overcoming Test Anxiety

If you struggle with test anxiety, we strongly encourage you to check out our recommendations for how you can overcome it. Test anxiety is a formidable foe, but it can be beaten, and we want to make sure you have the tools you need to defeat it.

Chapter 1: Solving Equations

1.1 Solving One-Step Equations

PRACTICE QUESTIONS

1. The equation $x - 5 = 12$ is given. What is the value of x?
    a. $x = 6$
    b. $x = 7$
    c. $x = 17$
    d. $x = 60$

2. The equation $x + 7 = 25$ is given. Solve for x.
    a. $x = 18$
    b. $x = 30$
    c. $x = 32$
    d. $x = 175$

3. What is the value of x in the equation $\frac{x}{5} = -6$?
    a. $x = -30$
    b. $x = -11$
    c. $x = 11$
    d. $x = 30$

4. The equation $-3x = 45$ is given. What is the value of x?
    a. $x = -135$
    b. $x = -15$
    c. $x = 15$
    d. $x = 135$

5. The equation $y - 14 = 63$ is given. What is the value of y?
    a. $y = 49$
    b. $y = 57$
    c. $y = 71$
    d. $y = 77$

6. What is the value of y in the equation $y + 27 = -89$?
    a. $y = -116$
    b. $y = -62$
    c. $y = 62$
    d. $y = 116$

7. The equation $\frac{y}{17} = 22$ is given. What is the value of y?
    a. $y = 5$
    b. $y = 39$
    c. $y = 110$
    d. $y = 374$

8. The equation $-15y = 375$ is given. Solve for y.
  a. $y = -5,625$
  b. $y = -25$
  c. $y = 25$
  d. $y = 5,625$

9. The equation $w - 57 = -279$ is given. Solve for w.
  a. $w = -336$
  b. $w = -222$
  c. $w = 222$
  d. $w = 336$

10. The equation $w + 109 = -73$ is given. What is the value of w?
  a. $w = -182$
  b. $w = -36$
  c. $w = 36$
  d. $w = 182$

11. The equation $\frac{w}{92} = 5$ is given. What is the value of w?
  a. $w = 18$
  b. $w = 87$
  c. $w = 97$
  d. $w = 460$

12. What is the value of w in the equation $-30w = 240$?
  a. $w = -7,200$
  b. $w = -8$
  c. $w = 210$
  d. $w = 60$

13. Kusha and Zola pick 37 mangoes from a tree. If Kusha picked 22 mangoes, how many mangoes did Zola pick?
  a. 15 mangoes
  b. 19 mangoes
  c. 55 mangoes
  d. 59 mangoes

14. Alex is taking a math test that requires one hour to complete. There are two sections of the test. Alex completes the first section in 41 minutes. Write a one-step equation to represent the time Alex spends on the two tests, then solve for the amount of time that Alex has left to complete the second section of the test.
  a. $60 = 14 + y$, 46 minutes for section two
  b. $60 = 41 + y$, 19 minutes for section two
  c. $40 = 61 + y$, 21 minutes for section two
  d. $41 = 60 + y$, 19 minutes for section two

15. Joan is selling raffle tickets at a basketball game for $8 dollars each. How many tickets does she need to sell in order to make $440? Write a one-step equation in order to solve for the number of tickets and then solve for the number of tickets.

   a. $440 = 55x$, 8 tickets need to be sold
   b. $8x = 440$, 65 tickets need to be sold
   c. $550 = 8x$, 50 tickets need to be sold
   d. $440 = 8x$, 55 tickets need to be sold

1.2 Solving Two-Step Equations

Practice Questions

1. Solve for the value of x that will make the following equation true: $2x + 4 = 8$.
    a. $x = 4$
    b. $x = 1$
    c. $x = 2$
    d. $x = 2.5$

2. Solve for the value of x that satisfies the equation $-4x - 3 = 13$.
    a. $x = 4$
    b. $x = 3$
    c. $x = -5$
    d. $x = -4$

3. Solve for the value of x that makes the following equation true: $10x + 40 = 100$.
    a. $x = 6$
    b. $x = 5$
    c. $x = 4$
    d. $x = 8$

4. Solve the following two-step equation for z: $6z - 2 = 10$.
    a. $z = 2$
    b. $z = 3$
    c. $z = 4$
    d. $z = 6$

5. Solve the following equation for x: $\frac{x}{5} - 3 = 2$.
    a. $x = 1$
    b. $x = 25$
    c. $x = 15$
    d. $x = 30$

6. Solve for n in the following equation: $-n + 5 = -17$.
    a. $n = 22$
    b. $n = -12$
    c. $n = -22$
    d. $n = 12$

7. Solve the following equation for the value of x that will make it true: $\frac{x}{3} - 7 = 11$.
    a. $x = \frac{7}{3}$
    b. $x = 77$
    c. $x = \frac{4}{3}$
    d. $x = 54$

8. Solve for x in the following two-step equation: $-3x - 2 = -1$.
   a. $x = \frac{1}{3}$
   b. $x = -1$
   c. $x = -\frac{1}{3}$
   d. $x = 3$

9. Solve for x in the equation $9x + 36 = 72$.
   a. $x = \frac{4}{3}$
   b. $x = 6$
   c. $x = 8$
   d. $x = 4$

10. Solve for x in the equation $-7x + 4 = -10$.
    a. $x = \frac{6}{7}$
    b. $x = -2$
    c. $x = -\frac{6}{7}$
    d. $x = 2$

11. Solve for x in the equation $8x + 1 = -21$.
    a. $x = \frac{5}{2}$
    b. $x = -\frac{11}{4}$
    c. $x = -3$
    d. $x = -4$

12. Solve for x given the equation $-\frac{3}{5}x + 8 = 23$.
    a. $x = -25$
    b. $x = \frac{14}{5}$
    c. $x = -9$
    d. $x = 9$

13. Danielle is interested in buying some blouses that are on sale for $6.00 each. She has a budget of $32, and she also has a $10-off coupon. Write an equation from the information given and determine how many blouses Danielle can buy.
    a. $6x = 22$, and Danielle can buy 3 blouses.
    b. $6x - 10 = 32$, and Danielle can buy 7 blouses.
    c. $10x - 6 = 32$, and Danielle can buy 3 blouses.
    d. $6x - 10 = 32$, and Danielle can buy 8 blouses.

14. Billy rides his bicycle 5 miles each morning that he works his paper route. One morning this week, he rode an extra mile to visit with his grandparents. At the end of the week, he had ridden 21 miles. How many mornings did he deliver papers? Support your answer with an equation.

   a. 3 mornings; $6x + 3 = 21$
   b. 4 mornings; $21x - 5 = 22$
   c. 4 mornings; $5x + 1 = 21$
   d. 4 mornings; $6x + 3 = 21$

15. Nora earns $4 per hour at her waitressing job and today received $29 in tips. From her shift today, she earned a total of $53 from both tips and hourly wages. Write an equation from the information given and determine how many hours Nora worked today.

   a. $4x + 29 = 53$; Nora worked 6 hours.
   b. $29x + 4 = 53$; Nora worked 2 hours.
   c. $53x - 29 = 4$; Nora worked 2 hours.
   d. $4x - 29 = 53$; Nora worked 9 hours.

1.3 Solving Equations Using the Distributive Property

PRACTICE QUESTIONS

1. Simplify the expression $3(2x - 8)$.
   a. $6x - 8$
   b. $2x - 24$
   c. $5x - 11$
   d. $6x - 24$

2. Simplify the expression $4(x + 13)$.
   a. $x + 52$
   b. $4x + 52$
   c. $5x + 52$
   d. $5x + 17$

3. Simplify the expression $6(3x + 9)$.
   a. $x + 54$
   b. $9x + 15$
   c. $18x + 54$
   d. $18x + 15$

4. Simplify the expression $5(4x - 7)$.
   a. $9x - 12$
   b. $9x - 35$
   c. $20x - 12$
   d. $20x - 35$

5. The equation $2(3y + 6) = 60$ is given. What is the value of y?
   a. $y = 8$
   b. $y = 12$
   c. $y = 16$
   d. $y = 24$

6. What is the value of y in the equation $8(y + 8) = -56$?
   a. $y = -15$
   b. $y = -5$
   c. $y = 5$
   d. $y = 15$

7. The equation $-3(4y - 9) = 303$ is given. What is the value of y?
   a. $y = -28$
   b. $y = -23$
   c. $y = 23$
   d. $y = 28$

8. The equation $4(-4y + 9) = 180$ is given. Solve for y.
   a. $y = -18$
   b. $y = -9$
   c. $y = 14$
   d. $y = 27$

9. The equation $5(2w + 4) = 8(w + 9)$ is given. Solve for w.
   a. $w = 5$
   b. $w = 13$
   c. $w = 26$
   d. $w = 46$

10. The equation $6(3w - 5) = 4(6w + 3)$ is given. What is the value of w?
    a. $w = -7$
    b. $w = -3$
    c. $w = 3$
    d. $w = 7$

11. The equation $-7(3w + 6) = -2(3w - 84)$ is given. What is the value of w?
    a. $w = -21$
    b. $w = -14$
    c. $w = 14$
    d. $w = 21$

12. Solve for the variable n.
$$6(n + 3) = -2(n + 31)$$
    a. $n = -10$
    b. $n = 5.6$
    c. $n = -4$
    d. $n = -5$

13. The length of a rectangular garden is $x + 5$ feet, and the width of the garden is x feet. What is the perimeter of the rectangular garden?
    a. $2x + 5$ feet
    b. $2x + 10$ feet
    c. $4x + 7$ feet
    d. $4x + 10$ feet

14. Tima and 4 of her friends go to the movies. The ticket for the movie costs x dollars. They each buy a small popcorn for y dollars and a small drink for z dollars. Which expression shows the total cost of Tima and her friends going to the movies?
    a. $4xyz$ dollars
    b. $5xyz$ dollars
    c. $4x + 4y + 4z$ dollars
    d. $5x + 5y + 5z$ dollars

15. Mr. Jonas buys 15 pens, pencils, and erasers that cost $1.15, $0.79, and $1.25 each, respectively. How much does Mr. Jonas spend in total on the supplies for his students?
   a. $11.85
   b. $17.25
   c. $18.75
   d. $47.85

1.4 Solving Equations with Variables on Both Sides

Practice Questions

1. Solve for the variable m.

$$9m + 4 = 7m + 12$$

a. $m = 4$
b. $m = 2$
c. $m = -4$
d. $m = 3.5$

2. Solve for the variable x.

$$2x - 7 = 5x + 8$$

a. $x = -5$
b. $x = 5.5$
c. $x = 2.5$
d. $x = 5$

3. The equation $2x + 3 = x + 16$ is given. What is the value of x?

a. $x = 8$
b. $x = 13$
c. $x = 19$
d. $x = 21$

4. The equation $4x - 8 = x + 19$ is given. Solve for x.

a. $x = 9$
b. $x = 10$
c. $x = 27$
d. $x = 30$

5. What is the value of x in the equation $\frac{5x}{7} = -18 + 2x$?

a. $x = -12$
b. $x = -7$
c. $x = 14$
d. $x = 36$

6. The equation $-6x + 7 = x + 56$ is given. What is the value of x?

a. $x = -9$
b. $x = -7$
c. $x = 7$
d. $x = 9$

7. The equation $-8y + 5 = -2y + 53$ is given. What is the value of y?

a. $y = -8$
b. $y = -6$
c. $y = 6$
d. $y = 8$

8. What is the value of y in the equation $3y - 12 = -5y + 28$?
    a. $y = 2$
    b. $y = 5$
    c. $y = 8$
    d. $y = 20$

9. The equation $-\frac{y}{2} - 26 = -3y + 49$ is given. What is the value of y?
    a. $y = 25$
    b. $y = 30$
    c. $y = 46$
    d. $y = 150$

10. The equation $-14y - 33 = -12y + 29$ is given. Solve for y.
    a. $y = -37$
    b. $y = -31$
    c. $y = 31$
    d. $y = 37$

11. The equation $6w + 16 = 4w - 30$ is given. Solve for w.
    a. $w = -23$
    b. $w = -7$
    c. $w = 7$
    d. $w = 23$

12. The equation $-7w - 52 = -3w + 20$ is given. What is the value of w?
    a. $w = -18$
    b. $w = -8$
    c. $w = 8$
    d. $w = 18$

13. Two pools at a swim club are being drained at the same time. Pool 1 has 2,100 gallons of water and is draining at 50 gallons per minute. Pool 2 has 3,600 gallons of water and is draining at 80 gallons per minute. After how many minutes, t, will the two pools have the same amount of water?
    a. 43 minutes
    b. 50 minutes
    c. 54 minutes
    d. 71 minutes

14. Car rental A charges a $15 daily fee and $0.20 per mile. Car rental B charges a $10 daily fee and $0.40 per mile. After how many miles, m, in one day will the cost of the two cars be the same?
    a. 8.3 miles
    b. 25 miles
    c. 41.7 miles
    d. 125 miles

15. A sign store charges $122 for a sign and $4 per letter to print on the sign. A second sign store charges $158 for a sign and $2.50 per letter to print on the sign. How many letters, p, would it take for the two signs to cost the same?

   a. 6 letters
   b. 14 letters
   c. 24 letters
   d. 43 letters

1.5 Simplifying Algebraic Expressions

Practice Questions

1. What is $(3m + n + 6) + (8m + 3n - 8)$ in its most simplified form?
   a. $5m + 3n - 2$
   b. $5m + 4n - 14$
   c. $11m + 4n - 2$
   d. $11m + 3n - 14$

2. Simplify the expression $(6m - 4n - 9) + (-2m - 3n + 2)$.
   a. $4m - n - 11$
   b. $4m - 7n - 7$
   c. $8m - n - 7$
   d. $8m - 7n - 11$

3. Which expression shows the most simplified form of the expression $(7m + 3n - 5) - (3m + 5n + 8)$?
   a. $4m - 2n - 13$
   b. $4m + 8n - 3$
   c. $11m - 2n - 3$
   d. $11m + 8n - 13$

4. Which shows the expression $(-4xy + 2x + 3) - (xy + x - 9)$ in its most simplified form?
   a. $-5xy + x + 12$
   b. $-5xy + 3x - 6$
   c. $-3xy + x + 12$
   d. $-3xy + 3x - 6$

5. What is $(y + x + 10) - (5y - 6x + 3)$ in its most simplified form?
   a. $-6y - 5x + 13$
   b. $-4y + 7x + 7$
   c. $4y - 5x + 7$
   d. $6y + 7x + 13$

6. Which shows the expression $(5x - 3y - 5) + (-2x - 8y - 6)$ in its most simplified form?
   a. $3x - 11y - 11$
   b. $3x - 5y - 1$
   c. $7x - 11y - 1$
   d. $7x - 5y - 11$

7. Which expression shows the most simplified form of the expression $(9y + 3xy + 8) + (-3y - xy - 2)$?
   a. $6y + 4xy + 10$
   b. $6y + 2xy + 6$
   c. $12y + 2xy + 10$
   d. $12y + 4xy + 6$

8. Simplify the expression $(7y - 4xy - 5) - (-3y - 2xy - 2)$.
    a. $4y - 6xy - 7$
    b. $4y - 2xy - 3$
    c. $10y - 6xy - 7$
    d. $10y - 2xy - 3$

9. Which expression shows the most simplified form of the expression $(-3u - 4v + 1) + (-5u + v + 4)$?
    a. $-8u - 5v + 3$
    b. $-8u - 3v + 5$
    c. $-2u - 5v + 5$
    d. $-2u - 3v + 3$

10. Which shows the expression $(-6w + 3vw - 9) - (-2w - 4vw + 6)$ in its most simplified form?
    a. $-8w - vw - 15$
    b. $-8w + 7vw - 3$
    c. $-4w - vw - 3$
    d. $-4w + 7vw - 15$

11. Which expression shows the most simplified form of the expression $(8w - 5vw + 2) + (-3w - 6vw - 7)$?
    a. $5w - 11vw - 5$
    b. $5w - vw - 9$
    c. $11w - 11vw - 5$
    d. $11w - vw - 9$

12. Simplify the following expression: $(14xy + 3z - 8) - (2xy + 5z - 4)$.
    a. $12xy + 2z + 4$
    b. $12x - 2xz - 4$
    c. $12xy - 2z + 1$
    d. $12xy - 2z - 4$

13. What is the perimeter of a rectangle with length $2x + 3y - 5$ meters and width $x + y - 2$ meters?
    a. $3x + 4y - 7$ meters
    b. $3x + 4y + 7$ meters
    c. $6x + 8y - 14$ meters
    d. $6x + 8y + 14$ meters

14. Amazing Bakery buys $10y + 2x + 5$ pounds of flour and uses $3y - 5x + 3$ pounds of flour for cakes. How many pounds of flour does the bakery have left?
    a. $7y - 3x + 8$ pounds
    b. $7y + 7x + 2$ pounds
    c. $13y - 3x + 2$ pounds
    d. $13y + 7x + 8$ pounds

15. Mary travels $5y + 3x + 12$ miles from home to school and $4y - 7x + 9$ miles from school to work. How many total miles does Mary travel from home to work?

   a. $y - 4x + 3$ miles
   b. $y + 4x + 21$ miles
   c. $9y - 4x + 21$ miles
   d. $9y + 4x + 3$ miles

1.6 Solving Equations by Combining Like Terms

Practice Questions

1. Solve for the variable x.
$$14x - 1 = 111$$
   a. $x = 9$
   b. $x = 8$
   c. $x = 7$
   d. $x = 6$

2. Solve for the variable x.
$$3x - 2.5x = 6.5 - 10.5$$
   a. $x = -6$
   b. $x = -8$
   c. $x = 8$
   d. $x = 4$

3. Solve for the variable t.
$$-5t + 3t + 1 = -14$$
   a. $t = 6$
   b. $t = 6.5$
   c. $t = 7$
   d. $t = 7.5$

4. Solve for the variable m.
$$3m - 6 + 9m = 30m + 9$$
   a. $m = -\frac{5}{6}$
   b. $m = \frac{5}{6}$
   c. $m = -\frac{4}{9}$
   d. $m = -\frac{3}{5}$

5. Solve for the variable x.
$$-7x + 10x - 12 = 3$$
   a. $x = 5.5$
   b. $x = 6$
   c. $x = 5$
   d. $x = 6.5$

6. Solve for the variable n.
$$-1 + 3n = -7 - 6n$$
   a. $n = 2$
   b. $n = 3$
   c. $n = \frac{2}{3}$
   d. $n = -\frac{2}{3}$

7. The terms in each list are all considered like terms except for which list?
   a. $2x, -7x, 2x$
   b. $5xy, 9xy, xy$
   c. $8x^2, 14x^2, -2x^2$
   d. $45x, 45y, 45z$

8. Solve for the variable y.
$$3y + 5 - 2y = 6y - 10$$
   a. $y = 3$
   b. $y = 5$
   c. $y = -2$
   d. $y = -7$

9. Solve for the variable k.
$$-2k - 3 = -2(2k + 1)$$
   a. $k = 3.5$
   b. $k = 2.5$
   c. $k = \frac{1}{2}$
   d. $k = \frac{4}{5}$

10. Solve for the variable x.
$$-2(-2x + 3) + 4x = -2(6 - 3x) + 3x$$
   a. $x = 3$
   b. $x = 5$
   c. $x = 8$
   d. $x = 6$

11. Solve for the variable y.
$$-\frac{1}{2}(10y - 2) + 3 = 14$$
   a. $y = -2$
   b. $y = 2$
   c. $y = \frac{1}{2}$
   d. $y = -\frac{1}{4}$

12. Solve for the variable n.
$$5(n + 3) - n = 2(n - 4) + 3$$
   a. $n = -10$
   b. $n = 8$
   c. $n = -13$
   d. $n = 9$

13. Ruby is working on solving equations in math class. Ruby's teacher puts the following problem on the board:
$$6 + 4w = 4w - 3 + 30w - 6$$
The teacher tells the class to simplify the equation by combining like terms before solving for w. What should Ruby's simplified equation be?
   a. $30w = -15$
   b. $38w = -15$
   c. $15 = 30w$
   d. $15 = 38w$

14. In 2010, the cost of a laptop was $(-4c^2 + 450)$ dollars. In 2020, the cost of a laptop was $(-8c^2 + 600)$ dollars. To represent the difference in laptop costs, Marcus subtracts the 2010 cost from the 2020 cost: $(-8c^2 + 600) - (-4c^2 + 450)$. Simplify Marcus's expression by applying the distributive property and combining like terms.
   a. $-12c^2 + 150$
   b. $-4c^2 + 150$
   c. $-12c^2 + 1,050$
   d. $-4c^2 + 1,050$

15. Eddie and his friends are going to a water park that charges $7 to enter the park and then $2 per ride. He purchases lunch at the park for $12. At the end of the day Eddie has spent $31. How many rides did he go on?
   a. 6 rides
   b. 4 rides
   c. 11 rides
   d. 8 rides

1.7 Solving Multi-Step Equations

Practice Questions

1. Solve for the variable x.

$$-3(4x + 3) + 4(6x + 1) = 43$$

a. $x = 7$
b. $x = 6$
c. $x = 5$
d. $x = 4$

2. Solve for the variable x.

$$3(x + 12) - x = 8$$

a. $x = 17$
b. $x = -14$
c. $x = 14$
d. $x = -17$

3. Solve for the variable x.

$$\frac{3}{10}(x + 2) = 12$$

a. $x = \frac{3}{8}$
b. $x = 21$
c. $x = 38$
d. $x = 3\frac{1}{2}$

4. Solve for the variable x.

$$-\frac{1}{2}(x + 2) + 3x = -1$$

a. $x = -3$
b. $x = -1$
c. $x = \frac{1}{2}$
d. $x = 0$

5. Solve for the variable y.

$$3y - 5 = -8(6 + 5y)$$

a. $y = \frac{1}{4}$
b. $y = -6$
c. $y = -1$
d. $y = 48$

6. Solve for the variable y.

$$2(y + 2) - 5 = 3(y + 1)$$

a. $y = -4$
b. $y = -5$
c. $y = \frac{1}{2}$
d. $y = 14$

13. Ruby is working on solving equations in math class. Ruby's teacher puts the following problem on the board:
$$6 + 4w = 4w - 3 + 30w - 6$$
The teacher tells the class to simplify the equation by combining like terms before solving for w. What should Ruby's simplified equation be?
   a. $30w = -15$
   b. $38w = -15$
   c. $15 = 30w$
   d. $15 = 38w$

14. In 2010, the cost of a laptop was $(-4c^2 + 450)$ dollars. In 2020, the cost of a laptop was $(-8c^2 + 600)$ dollars. To represent the difference in laptop costs, Marcus subtracts the 2010 cost from the 2020 cost: $(-8c^2 + 600) - (-4c^2 + 450)$. Simplify Marcus's expression by applying the distributive property and combining like terms.
   a. $-12c^2 + 150$
   b. $-4c^2 + 150$
   c. $-12c^2 + 1,050$
   d. $-4c^2 + 1,050$

15. Eddie and his friends are going to a water park that charges $7 to enter the park and then $2 per ride. He purchases lunch at the park for $12. At the end of the day Eddie has spent $31. How many rides did he go on?
   a. 6 rides
   b. 4 rides
   c. 11 rides
   d. 8 rides

1.7 Solving Multi-Step Equations

Practice Questions

1. Solve for the variable x.

$$-3(4x + 3) + 4(6x + 1) = 43$$

   a. $x = 7$
   b. $x = 6$
   c. $x = 5$
   d. $x = 4$

2. Solve for the variable x.

$$3(x + 12) - x = 8$$

   a. $x = 17$
   b. $x = -14$
   c. $x = 14$
   d. $x = -17$

3. Solve for the variable x.

$$\frac{3}{10}(x + 2) = 12$$

   a. $x = \frac{3}{8}$
   b. $x = 21$
   c. $x = 38$
   d. $x = 3\frac{1}{2}$

4. Solve for the variable x.

$$-\frac{1}{2}(x + 2) + 3x = -1$$

   a. $x = -3$
   b. $x = -1$
   c. $x = \frac{1}{2}$
   d. $x = 0$

5. Solve for the variable y.

$$3y - 5 = -8(6 + 5y)$$

   a. $y = \frac{1}{4}$
   b. $y = -6$
   c. $y = -1$
   d. $y = 48$

6. Solve for the variable y.

$$2(y + 2) - 5 = 3(y + 1)$$

   a. $y = -4$
   b. $y = -5$
   c. $y = \frac{1}{2}$
   d. $y = 14$

7. Solve for the variable y.

$$-6y + 3 - 7 = 20$$

a. $y = -4$
b. $y = 6$
c. $y = 4$
d. $y = 5$

8. Solve for the variable y.

$$5(y - 3) = 6y + 10$$

a. $y = -13$
b. $y = 18$
c. $y = -25$
d. $y = 25.5$

9. Solve for the variable n.

$$-3(n - 6) + 4(n + 1) = 7n - 10$$

a. $n = 7$
b. $n = 12$
c. $n = \frac{3}{16}$
d. $n = \frac{16}{3}$

10. Solve for the variable n.

$$\frac{n + 4}{3} = 2$$

a. $n = 2$
b. $n = 10$
c. $n = 1$
d. $n = 9$

11. Solve for the variable n.

$$5(1 + 4n) + 2n = 27$$

a. $n = 11$
b. $n = 1\frac{1}{2}$
c. $n = 1$
d. $n = 1.1$

12. Solve for the variable n.

$$2(5n - 8) = -6(n - 8)$$

a. $n = 3.5$
b. $n = 4$
c. $n = 2$
d. $n = -8$

13. If the triangle and the square have the same perimeter, then what is the value of x? Set up a multi-step equation to help you solve for x.

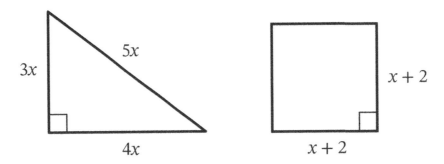

   a. $x = 4$
   b. $x = 1$
   c. $x = 2$
   d. $x = 1.5$

14. Mrs. Rodríguez has 8 busses available for a class field trip. She has 315 students attending the field trip, and 11 of these students say they don't need to ride the bus because they can get a ride from a family member. Set up an equation to solve for the number of students that will be on each bus. Each bus will carry the same number of students.

   a. 38 students per bus
   b. 26 students per bus
   c. 21 students per bus
   d. 33 students per bus

15. John wants to calculate the number of minutes he spent on the phone this month. He pays $30 per month plus $0.07 per minute, and his phone bill this month was $59.40. Set up a multi-step equation to solve for the number of minutes John spent on the phone this month.

   a. 310 minutes
   b. 200 minutes
   c. 505 minutes
   d. 420 minutes

1.8 Solving Equations Involving Absolute Value

PRACTICE QUESTIONS

1. Solve for the variable x given the equation $|x| = 15$.
   a. $x = 15$ or $x = -15$
   b. $x = 15$
   c. $x = 15$ or $x = 51$
   d. $x = 0$ or $x = -15$

2. Solve for the variable x in the equation $|x| = 7$.
   a. $x = 3$ or $x = -3$
   b. $x = 7$ or $x = -7$
   c. $x = 7$ or $x = -77$
   d. $x = 7$ or $x = 0$

3. The equation $|x| - 3 = 7$ is given. What is the value of x?
   a. $x = 3$ or $x = -3$
   b. $x = 4$ or $x = -4$
   c. $x = 10$ or $x = -10$
   d. $x = 11$ or $x = -11$

4. The equation $|x| + 5 = 9$ is given. Solve for x.
   a. $x = 4$ or $x = -4$
   b. $x = 5$ or $x = -5$
   c. $x = 13$ or $x = -13$
   d. $x = 14$ or $x = -14$

5. Solve for the variable x given the equation $|2x - 4| = 4$.
   a. $x = 0$
   b. $x = 4$
   c. $x = 4$ and $x = 0$
   d. $x = 3$ and $x = 0$

6. Solve for the variable x given the equation $|2x + 6| = 10$.
   a. $x = -2$ and $x = 8$
   b. $x = -8$
   c. $x = 2$
   d. $x = 2$ and $x = -8$

7. What is the value of x in the equation $4|x + 2| - 6 = 18$?
   a. $x = 1$ or $x = -5$
   b. $x = 4$ or $x = -8$
   c. $x = 14$ or $x = -18$
   d. $x = 18$ or $x = -22$

8. The equation $|2x - 8| + 3 = 21$ is given. What is the value of x?
   a. $x = 9$ or $x = -1$
   b. $x = 12$ or $x = -4$
   c. $x = 13$ or $x = -5$
   d. $x = 16$ or $x = -8$

9. The equation $6|3y - 9| = 90$ is given. What is the value of y?
   a. $y = 2$ or $y = -8$
   b. $y = 3$ or $y = -3$
   c. $y = 8$ or $y = -2$
   d. $y = 35$ or $y = -29$

10. What is the value of y in the equation $2|2y - 4| - 3 = 17$?
    a. $y = 3$ or $y = -3$
    b. $y = 7$ or $y = -3$
    c. $y = 9$ or $y = -5$
    d. $y = 12$ or $y = -8$

11. The equation $|6y + 3| - 12 = 9$ is given. What is the value of y?
    a. $y = -3$ or $y = -4$
    b. $y = 1$ or $y = -1$
    c. $y = 3$ or $y = -4$
    d. $y = 4$ or $y = -4$

12. The equation $3|y - 5| - 16 = 11$ is given. Solve for y.
    a. $y = -14$ or $y = 4$
    b. $y = 14$ or $y = -4$
    c. $y = 20$ or $y = -10$
    d. $y = 35$ or $y = -25$

13. A cereal company has a product with 700 g of granola. The scale accepts any bag that is within 20 g of the required weight. What are the maximum and minimum weights that the scale will accept?
    a. Max: 680 g, Min: 660 g
    b. Max: 720 g, Min: 680 g
    c. Max: 740 g, Min: 700 g
    d. Max: 760 g, Min: 720 g

14. James scores an average of 38 points per game, with the game points being within 7 points of his game average. What is the lowest possible score for James during a game?
    a. 19
    b. 26
    c. 31
    d. 45

15. A real estate company is selling a house that is worth $280,000. They are willing to accept any offer that is within 3% of the value of the house. What is the least amount of money that the real estate company can accept for the house?

   a. $196,400
   b. $271,600
   c. $288,400
   d. $364,600

1.9 Solving Formulas for Specific Variables

PRACTICE QUESTIONS

1. Given the formula $e = mc^2$, solve for m.
   a. $m = \frac{c^2}{e}$
   b. $m = \frac{e}{c^2}$
   c. $m = \frac{e}{c}$
   d. $m = ec^2$

2. The area of a circle can be calculated using the formula $A = \pi r^2$. Given this formula, solve for the circle's radius, r.
   a. $r = \frac{A}{\pi}$
   b. $r = \frac{A}{\pi^2}$
   c. $r = A\sqrt{\frac{1}{\pi}}$
   d. $r = \sqrt{\frac{A}{\pi}}$

3. From the Pythagorean theorem, $a^2 + b^2 = c^2$, solve for c.
   a. $c = a + b$
   b. $c = a^2 + b^2$
   c. $c = \sqrt{a^2} + \sqrt{b^2}$
   d. $c = \sqrt{a^2 + b^2}$

4. From the Pythagorean theorem, $a^2 + b^2 = c^2$, solve for b.
   a. $b = c - a$
   b. $b = \sqrt{c^2 - a^2}$
   c. $b = c^2 - a^2$
   d. $b = \sqrt{c^2} - \sqrt{a^2}$

5. Given the equation $4x + 2y - 8 = 4z$, solve for the variable y.
   a. $y = -4x + 4z + 4$
   b. $y = -2x - 2z + 8$
   c. $y = -2x + 2z + 4$
   d. $y = -4x + 4z + 8$

6. In physics, the term momentum refers to how difficult it will be to stop a moving object. Momentum is calculated using the formula $p = mv$, where p is momentum, m is the object's mass, and v is the object's velocity or speed. Rearrange this formula to solve for v.
   a. $v = mp$
   b. $v = \frac{m}{p}$
   c. $v = p - m$
   d. $v = \frac{p}{m}$

7. For any triangle, the sum of its three interior angles must equal 180°. This fact can be written into a formula as $A + B + C = 180°$. Rearrange this formula to get B by itself.

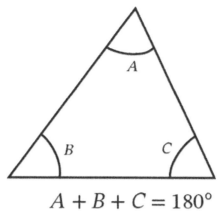

$$A + B + C = 180°$$

   a. $B = 180° + A + C$
   b. $B = 180° - A - C$
   c. $B = 180° - AC$
   d. $B = AC - 180$

8. In thermodynamics, scientists study heat and other forms of energy and how they move. One important formula used in thermodynamics is $Q = mc\Delta T$, where an object's mass m, specific heat c, and change in temperature ΔT are multiplied together to calculate heat transfer Q. Rearrange the formula to solve for ΔT.

   a. $\Delta T = \frac{Q}{mc}$
   b. $\Delta T = \frac{mc}{Q}$
   c. $\Delta T = mcQ$
   d. $\Delta T = Q - mc$

9. Use the formula $Q = mc\Delta T$ to solve for m.

   a. $m = Qc\Delta T$
   b. $m = Q - c - \Delta T$
   c. $m = \frac{cQ}{\Delta T}$
   d. $m = \frac{Q}{c\Delta T}$

10. The volume of a sphere can be found using the formula $V = \frac{4}{3}\pi r^3$. Solve for r.

   a. $r = \sqrt[3]{\frac{3}{4\pi}V}$
   b. $r = \sqrt[3]{\frac{4\pi}{3}V}$
   c. $r = V\sqrt[3]{\frac{3}{4\pi}}$
   d. $r = \frac{\sqrt[3]{V}}{\frac{4}{3}\pi}$

11. The formula for converting Fahrenheit to Celsius is $C = \frac{5}{9}(F - 32)$. Derive the formula for converting Celsius to Fahrenheit from this formula by solving for F.
   a. $F = \frac{9}{5}(C + 32)$
   b. $F = \frac{9}{5}C + 32$
   c. $F = \frac{5}{9}C + 32$
   d. $F = \frac{9}{5}C - 32$

12. One of the Pythagorean identities in trigonometry is $1 + tan^2\theta = sec^2\theta$. From this identity, solve for $tan^2\theta$.
   a. $tan^2\theta = sec\theta - 1$
   b. $tan^2\theta = sec^2\theta + 1$
   c. $tan^2\theta = \sqrt{sec\theta} - \sqrt{1}$
   d. $tan^2\theta = sec^2\theta - 1$

13. Scientists are sometimes interested in the densities of varying objects, which is the ratio of an object's mass to its size. This relationship is given in the formula $\rho = \frac{m}{V}$, where the Greek letter ρ (pronounced "row") is equal to the object's mass divided by the object's volume. A university physics lab has a block of copper with density $\rho = 8.96 \text{ g/cm}^3$ and volume $V = 10 \text{ cm}^3$. Rearrange the density formula to solve for m, and then determine the copper block's mass.
   a. $m = 0.896$ g
   b. $m = 1.116$ g
   c. $m = 89.6$ g
   d. $m = 18.96$ g

14. Morgan recently opened a savings account and put $200 ($p$) in it. Her bank pays interest at the rate of 2.5% (r) each year. If Morgan wants to earn $10 ($i$) in interest at this bank without adding or removing money from the account, how long will she have to wait? (Hint: rearrange the formula $i = prt$ to solve for t, then substitute in the known values of i, p, and r.)
   a. $t = 4$ years
   b. $t = 50$ years
   c. $t = \frac{1}{2}$ years
   d. $t = 2$ years

15. Mr. Ledger is an accountant for the Make It Count accounting firm, and he has the following financial information (in thousands) for this fiscal year:

$$\text{Total assets } (TA) = 850$$
$$\text{Total liabilities } (TL) = 425$$
$$\text{Common stock } (CS) = 175$$
$$\text{Retained Earnings } (RE) = ???$$

Mr. Ledger is still waiting to receive final information about this year's retained earnings, but he can determine that value by using the accounting equation: $TA = TL + CS + RE$. Rearrange this formula and solve for the value of RE.

  a. $RE = 250$
  b. $RE = 600$
  c. $RE = 1,100$
  d. $RE = 1,450$

1.10 Writing and Solving Equations Involving Direct Variation

PRACTICE QUESTIONS

1. Write a direct variation equation where $k = 5$.
    a. $5 = yx$
    b. $y = 5k$
    c. $y = 5x$
    d. $x = 5y$

2. Determine which of the following equations is an example of a direct variation equation.
    a. $y = 2x + 1$
    b. $y = 4x - 4$
    c. $y = 12x$
    d. $y = -3x + \frac{2}{3}$

3. Write a direct variation equation where $k = \frac{3}{5}$.
    a. $y = \frac{3}{5} + x$
    b. $x = \frac{3}{5}y$
    c. $y = \frac{3}{5}x$
    d. $y = \frac{3}{5x}$

4. Which of the following equations is not an example of direct variation?
    a. $y = -10x$
    b. $y = x$
    c. $y = \frac{1}{2}x$
    d. $y = 2x + 2$

5. Write a direct variation equation where $k = -3$, and evaluate the equation when $x = -1$.
    a. $y = 3x, y = -3$
    b. $y = -3x, y = 3$
    c. $y = 3x, y = 3$
    d. $y = -3x, y = -3$

6. Write a direct variation equation where $k = 12$, and evaluate the equation at $x = 5$.
    a. $y = 12x, y = 60$
    b. $y = 12 + x, y = 17$
    c. $y = 12x + 12, y = 72$
    d. $y = \frac{12}{x}, y = \frac{12}{5}$

7. For problems involving direct variation, we have the equation $y = kx$. What is the formula used for direct variation problems when you are given x and y and want to solve for k?
    a. $x = yk$
    b. $k = x + y$
    c. $k = \frac{y}{x}$
    d. $x = \frac{k}{y}$

8. Determine whether the information given in the table below represents an example of direct variation. If it does, find the value of k.

x	y
1	-5
3	-15
5	-25

a. This data does not represent an example of direct variation.
b. This data represents an example of direct variation and $k = 5$.
c. This data represents an example of direct variation and $k = -\frac{1}{5}$.
d. This data represents an example of direct variation and $k = -5$.

9. Determine whether the information given in the table below represents an example of direct variation. If it does, find the value of k.

x	y
10	20
12	36
14	42

a. This data does not represent an example of direct variation.
b. This data represents an example of direct variation and $k = 2.5$.
c. This data represents an example of direct variation and $k = 3$.
d. This data represents an example of direct variation and $k = 2$.

10. Determine whether the information given in the table below represents an example of direct variation. If it does, write the direct variation equation that corresponds to the data.

x	y
3	-2
6	-4
9	-6

a. This data does not represent an example of direct variation.
b. This data represents an example of direct variation and $y = -\frac{3}{2}x$.
c. This data represents an example of direct variation and $y = -\frac{2}{3}x$.
d. This data represents an example of direct variation $y = \frac{3}{2} + x$.

11. Determine whether the information given in the table below represents an example of direct variation. If it does, find the value of k.

x	y
−4	16
−2	8
0	0
2	−8

a. This data does not represent an example of direct variation.
b. This data represents an example of direct variation and $k = -\frac{1}{4}$.
c. This data represents an example of direct variation and $k = 4$.
d. This data represents an example of direct variation and $k = -4$.

12. Fill in the following table, given that the data are an example of direct variation and $k = 2$.

x	y
	−9
	−2
	5
	12

a. $x = -18, -4, 10, 24$
b. $x = -4.5, -1, 2.5, 6$
c. $x = -9, -2, 5, 12$
d. $x = 4.5, 1, -2.5, -6$

13. On average, adults typically consume about 2,000 calories per day. Let x represent "number of days" and write a direct variation equation describing how many calories an adult consumes on average given a certain number of days.

a. $y = 2{,}000x$
b. $y = 2{,}000 + x$
c. $y = 2x + 1{,}000$
d. $y = \frac{x}{2{,}000}$

14. Chase is driving 60 mph and still has 2.5 hours of driving on his trip. Using his speed, write a direct variation equation and solve to determine how many miles he has left to drive.

a. $y = 60x$; 125 miles
b. $y = 60x$; 150 miles
c. $y = 25x$; 180 miles
d. $y = 2.5x$; 125 miles

15. Casey is getting ready to bake cookies for a bake sale. If the recipe calls for half a cup of flour for a dozen cookies, which of the following is a direct variation equation that represents how many cups of flour she will need for each dozen cookies she bakes?

a. $y = \frac{1}{2}x$
b. $y = 2x$
c. $y = x$
d. $y = \frac{1}{3}x$

Chapter 2: Functions and Sequences

2.1 Evaluating Functions

PRACTICE QUESTIONS

1. The function $f(x) = -3x - 8$ is given. What is $f(5)$?
    a. $f(5) = -23$
    b. $f(5) = -7$
    c. $f(5) = 7$
    d. $f(5) = 23$

2. The function $f(x) = -5x + 18$ is given. What is $f(-4)$?
    a. $f(-4) = -38$
    b. $f(-4) = -2$
    c. $f(-4) = 2$
    d. $f(-4) = 38$

3. The function $f(x) = 4x - 16$ is given. What is $f(9)$?
    a. $f(9) = 3$
    b. $f(9) = 20$
    c. $f(9) = 29$
    d. $f(9) = 52$

4. The function $g(x) = 7x + 25$ is given. What is $g(-10)$?
    a. $g(-10) = -45$
    b. $g(-10) = -42$
    c. $g(-10) = 42$
    d. $g(-10) = 45$

5. The function $g(x) = 5x - 33$ is given. What is $g(-6)$?
    a. $g(-6) = -63$
    b. $g(-6) = -44$
    c. $g(-6) = 44$
    d. $g(-6) = 63$

6. The function $g(x) = 9x + 26$ is given. What is $g(-7)$?
    a. $g(-7) = -90$
    b. $g(-7) = -37$
    c. $g(-7) = 37$
    d. $g(-7) = 90$

7. The function $g(x) = -10x + 31$ is given. What is $g(12)$?
    a. $g(12) = -151$
    b. $g(12) = -89$
    c. $g(12) = 89$
    d. $g(12) = 151$

8. Evaluate the function $f(x) = 3(x+8) - 16$ when $x = 6$.
    a. $f(6) = 1$
    b. $f(6) = 298$
    c. $f(6) = 26$
    d. $f(6) = 34$

9. Evaluate the function $f(x) = x^2 + 2x + 4$ when $x = 10$.
    a. $f(10) = 124$
    b. $f(10) = 44$
    c. $f(10) = 126$
    d. $f(10) = 234$

10. If $f(x) = x^4 + 2x^3$, find the value of $f(3) + 2f(4)$.
    a. 2,519
    b. 521
    c. 519
    d. 903

11. Michelle works at an appliance shop. She makes $15 per hour, plus an additional $30 for each sale she makes. On Tuesday, Michelle made 5 sales, so she earned an extra $150. Michelle's total pay for Tuesday can be represented by the function $f(x) = 15x + 150$, where x represents the number of hours worked. How much did Michelle earn altogether if she worked for 8 hours on Tuesday?
    a. $173
    b. $260
    c. $308
    d. $270

12. A gym membership costs $15 per week. The total cost of the membership is represented by the function $f(x) = 15x$, with x representing the number of weeks in the membership. The gym also rents out lockers for an additional $3 per week. The total cost of a membership with a locker rental is represented by the function $g(x) = 18x$. What is the difference between the cost of a 20-week membership with a locker rental and a 20-week membership without a locker rental?
    a. $40
    b. $60
    c. $30
    d. $20

13. When traveling, Maya gets paid a daily per diem of $38 for food and $0.58 per mile she travels using her car. If Maya travels to a city that is 75 miles away, what is her pay for traveling to and from the city in one day?
    a. $81.50
    b. $119.50
    c. $125.00
    d. $163.00

14. A kayak rental company charges $9 per hour and a $15 flat daily fee to rent a kayak. How much would it cost to rent a kayak for 5 hours?

 a. $24
 b. $60
 c. $84
 d. $136

15. A gym charges a one-time registration fee of $50 and $6 per week for membership. How much would someone pay for a 12-week gym membership fee?

 a. $56
 b. $68
 c. $122
 d. $606

2.2 Domain and Range of Linear Functions

PRACTICE QUESTIONS

1. Determine the domain and range of the linear function graphed below.

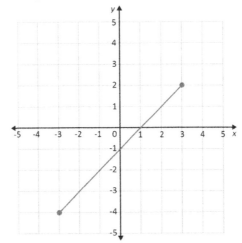

   a. Domain: $-3 < y < 3$
      Range: $-4 < x < 2$
   b. Domain: $-3 \leq x \leq 3$
      Range: $-4 \leq y \leq 2$
   c. Domain: $-3 \leq x \leq 3$
      Range: $-4 \leq y \leq 2$
   d. Domain: $-4 \leq y \leq 2$
      Range: $-3 \leq x \leq 3$

2. Determine the domain and range of the linear function graphed below.

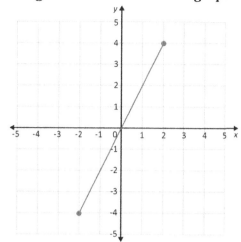

a. Domain: $-2 \leq y < 2$
   Range: $-4 < x \leq 4$
b. Domain: $-2 \leq x \leq 2$
   Range: $-4 \leq y \leq 4$
c. Domain: $-4 < y < 4$
   Range: $-2 < x < 2$
d. Domain: $-2 < x < 2$
   Range: $-4 < y < 4$

3. Determine the domain and range of the linear function graphed below.

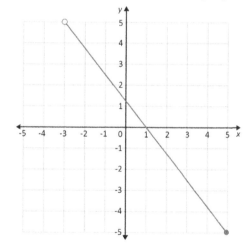

a. Domain: $-3 < x \leq 5$
   Range: $-5 \leq y < 5$
b. Domain: $-5 \leq y < 5$
   Range: $-3 < x \leq 5$
c. Domain: $-3 \leq x < 5$
   Range: $-5 < y \leq 5$
d. Domain: $-3 \leq y \leq 5$
   Range: $-5 \leq x \leq 5$

4. Determine the domain and range of the linear function graphed below.

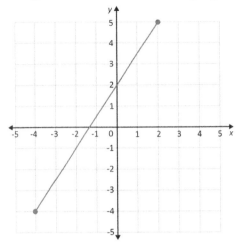

a. Domain: $-4 \leq y < 2$
   Range: $-4 < x \leq 5$
b. Domain: $-4 < x < 2$
   Range: $-4 < y < 5$
c. Domain: $-4 \leq y \leq 5$
   Range: $-4 \leq x \leq 2$
d. Domain: $-4 \leq x \leq 2$
   Range: $-4 \leq y \leq 5$

5. Here is a table of values that represents a linear function. What is the domain of the linear function?

x	y
-1	-8
2	-2
5	4
7	8

a. $\{-1,2,5,7\}$
b. $\{-8,-2,4,8\}$
c. $\{-8,-1,-2\}$
d. $\{2,4,5,7,8\}$

6. Here is a table of values that represents a linear function. What is the range of the linear function?

x	y
-3	0
-1	2
2	5
4	7

a. $\{-3,-1,2,4\}$
b. $\{0,2,5,7\}$
c. $\{-1,-3\}$
d. $\{0,2,2,4,5,7\}$

7. The graph of the linear function shows the cost of grapes as a function of the amount of grapes purchased in a grocery store. What is the domain of the linear function?

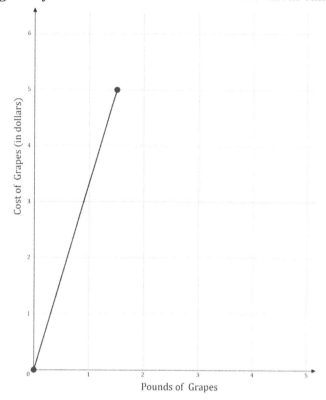

a. $0 \leq x \leq 1.5$
b. $0 \leq x \leq 5$
c. $0 \leq x \leq 6$
d. $1.5 \leq x \leq 5$

8. Use the linear function $f(x) = 4x + 1$ to determine a correct set of possible domain and range values.

a. D: $\{2, 4, 6, 8\}$, R: $\{-3, 5, 7, 8\}$
b. D: $\{1, 2, 3, 4\}$, R: $\{-5, -9, -13, -17\}$
c. D: $\{1, 2, 3, 4\}$, R: $\{5, 9, 13, 17\}$
d. D: $\{1, 3, 5, 7\}$, R: $\{5, 9, 11, 17\}$

9. Which domain and range values would satisfy the linear function $f(x) = -3$?

a. (1,3), (0,0), (1,4)
b. (-1,3), (0,0), (1,-3)
c. (-1,2), (0,5), (1,-3)
d. (1,3), (0,0), (1,9)

10. If the domain of the linear function $f(x) = 0.5x + 2$ is D: $\{-1, 0, 1, 2\}$, what is the range of the function?

a. R: $\{1.5, 2.5, 3.5, 4.5\}$
b. R: $\{-1.5, 2, -2.5, 3\}$
c. R: $\{1, 2, 3, 4\}$
d. R: $\{1.5, 2, 2.5, 3\}$

11. A bowling alley charges 8 dollars for use of the lane, plus an additional 3 dollars for each shoe rental. This relationship can be represented with the function $y = 3x + 8$. In this equation, x stands for the pairs of shoes rented, and y stands for the total cost to bowl. Determine reasonable domain and range values for this scenario.
    a. Domain: $\{0, 1, 2, 3, 4\}$
       Range: $\{0, 10, 12, 14, 16\}$
    b. Domain: $\{0, 1, 2, 3, 4\}$
       Range: $\{8, 11, 14, 17, 20\}$
    c. Domain: $\{-8, -11, -14, -17, -20\}$
       Range: $\{0, 1, 2, 3, 4\}$
    d. Domain: $\{0, -1, -2, -3, -4\}$
       Range: $\{6, 11, 13, 15, 17\}$

12. Hannah gets paid $12 an hour to babysit. She can use the equation $y = 12x$ to determine her pay for each babysitting job, where x is the number of hours she babysits and y is her total pay for that job. If Hannah can only work 6 hours a day, what is the range of the function $y = 12x$?
    a. $0 \leq y \leq 6$
    b. $0 \leq y \leq 72$
    c. $6 \leq y \leq 12$
    d. $6 \leq y \leq 72$

13. Sam gets paid $9 an hour working at Soccer Stop. He uses the equation $y = 9x$ to calculate his weekly pay, where x is the number of hours he worked and y is his total pay. If Sam's paycheck is always between $135 and $198, what is the domain of the function $y = 9x$?
    a. $15 \leq x \leq 22$
    b. $15 \leq x \leq 135$
    c. $22 \leq x \leq 198$
    d. $135 \leq x \leq 198$

14. An online video game rental company charges a monthly membership fee of $15 and $2 for each game that is rented. Dexter uses the linear function $y = 2x + 15$ to calculate his monthly bill, where x is the number of games he rents in one month and y is his total monthly bill. What is the range of the linear function if Dexter can only afford to rent a maximum of 8 games per month?
    a. $0 \leq y \leq 15$
    b. $0 \leq y \leq 31$
    c. $15 \leq y \leq 31$
    d. $17 \leq y \leq 31$

15. An online clothing rental company charges a monthly membership fee of $20 and $5 for each piece of clothing that is rented. Mike uses the linear function $y = 5x + 20$ to calculate his monthly bill, where x is the total pieces of clothing that he rented for the month and y is his monthly bill. What is the domain of the linear function if Mike's monthly bill is between $80 and $95, inclusive?

   a. {12}
   b. {15}
   c. {12,13,14,15}
   d. {13,14,15,16}

2.3 Manipulating Linear Functions

PRACTICE QUESTIONS

1. Which function represents the graph of $f(x) = x$ after it is stretched vertically by a factor of 3?
   a. $g(x) = 3f(x)$
   b. $g(x) = \frac{1}{3}f(x)$
   c. $g(x) = f(x) + 3$
   d. $g(x) = f(x + 3)$

2. Which function represents the graph of $f(x) = x$ after it is shifted 2 units up?
   a. $g(x) = f(x + 2)$
   b. $g(x) = f(x) + 2$
   c. $g(x) = f(x) - 2$
   d. $g(x) = f(x - 2)$

3. Which function represents the graph of $f(x) = x$ after it is shifted four units to the right?
   a. $g(x) = f(x + 4)$
   b. $g(x) = f(x - 4)$
   c. $g(x) = f(x) + 4$
   d. $g(x) = f(x) - 4$

4. Which function represents the graph of $f(x) = x$ after it is compressed horizontally by a factor of $\frac{1}{3}$?
   a. $g(x) = f(3x)$
   b. $g(x) = 3f(x)$
   c. $g(x) = f\left(\frac{1}{3}x\right)$
   d. $g(x) = \frac{1}{3}f(x)$

5. The coordinate plane below shows the graph of the parent function $f(x) = x$ (shown as a solid line), and a manipulation in which the parent function is compressed vertically by a factor of $\frac{1}{2}$ (shown as a dashed line). Which function represents this dilation?

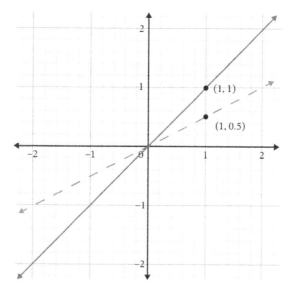

a. $g(x) = f(x) + \frac{1}{2}$
b. $g(x) = f\left(\frac{1}{2}x\right)$
c. $g(x) = \frac{1}{2}f(x)$
d. $g(x) = 2f(x)$

6. Which function represents the graph of $f(x) = x$ after it is shifted 5 units down?
a. $g(x) = f(x + 5)$
b. $g(x) = f(x) + 5$
c. $g(x) = f(x) - 5$
d. $g(x) = f(x - 5)$

7. Which function represents the graph of $f(x) = x$ after it is shifted five units to the left?
a. $g(x) = f(x) + 5$
b. $g(x) = f(x + 5)$
c. $g(x) = f(x) - 5$
d. $g(x) = f(x - 5)$

8. The graph of $f(x) = x$ is translated vertically 3 units up to create the graph of $g(x)$. What is the function of the translated graph?
a. $g(x) = f(x) + 3$
b. $g(x) = f(x) - 3$
c. $g(x) = f(x + 3)$
d. $g(x) = f(x - 3)$

9. The graph of $f(x) = x$ is translated vertically 4 units down to create the graph of $g(x)$. What is the function of the translated graph?

   a. $g(x) = f(x) + 4$
   b. $g(x) = f(x) - 4$
   c. $g(x) = f(x + 4)$
   d. $g(x) = f(x - 4)$

10. The graph of $f(x) = x$ is translated horizontally to the right by 2 units to create the graph of $g(x)$. What is the function of the translated graph?

    a. $g(x) = f(x) + 2$
    b. $g(x) = f(x) - 2$
    c. $g(x) = f(x + 2)$
    d. $g(x) = f(x - 2)$

11. The graph of $f(x) = x$ is translated horizontally to the left 5 units to create the graph of $g(x)$. What is the function of the translated graph?

    a. $g(x) = f(x) + 5$
    b. $g(x) = f(x) - 5$
    c. $g(x) = f(x + 5)$
    d. $g(x) = f(x - 5)$

12. The graph of $f(x) = x$ is stretched vertically by a factor of p to create the graph of $g(x)$. What is the function of the stretched graph?

    a. $g(x) = f(x) + p$
    b. $g(x) = f(x) - p$
    c. $g(x) = pf(x)$
    d. $g(x) = \frac{f(x)}{p}$

13. A skating rink rents ice skates for $1 per hour. The function $f(x) = x$ represents the hourly rental fee for x hours. The skating rink decides to add a $6 fee per rental to cover cleaning and maintenance expenses. The function that represents the new cost for skate rental is $g(x) = f(x) + 6$. How does the graph of this transformation compare to the graph of the parent function, $f(x)$?

    a. The graph of $g(x)$ is a translation of $f(x)$, with a vertical shift up 6 units.
    b. The graph of $g(x)$ is a translation of $f(x)$, with a vertical shift down 6 units.
    c. The graph of $g(x)$ is a translation of $f(x)$, with a horizontal shift right 6 units.
    d. The graph of $g(x)$ is a dilation of $f(x)$, vertically stretched by a factor of 6.

14. A law firm wants to create a smaller version of their logo to use on their letterhead. The edge of the original logo, represented by the function $f(x) = x$, needs to be compressed horizontally by a factor of $\frac{1}{4}$. Which function represents this dilation?

    a. $g(x) = \frac{1}{4}f(x)$
    b. $g(x) = 4f(x)$
    c. $g(x) = f\left(\frac{1}{4}x\right)$
    d. $g(x) = f(4x)$

15. Alex is a graphic designer and is working on creating a website. He needs to manipulate a linear image so it fits on the webpage. The original image, represented by the function $f(x) = x$, needs to be compressed vertically by a factor of $\frac{1}{4}$. Which function represents this dilation?

a. $g(x) = f(4x)$
b. $g(x) = 4f(x)$
c. $g(x) = f\left(\frac{1}{4}x\right)$
d. $g(x) = \frac{1}{4}f(x)$

2.4 Writing Functions for Real-World Problems

PRACTICE QUESTIONS

1. A car rental company charges a daily fee of $25 and $0.45 per mile driven. What is the total rental charge for one day if 150 miles was driven by the car?

   a. $67.50
   b. $92.50
   c. $161.25
   d. $175.45

2. A boat rental company charges a $15 daily fee and $4 per hour for renting a boat. Which equation can be used to calculate the total one-day rental fee, y, for renting the boat for x hours?

   a. $y = 4x + 15$
   b. $y = 15x + 4$
   c. $x = 4y + 15$
   d. $x = 15y + 4$

3. If purchasing a pound of apples costs $1.65, how much would 7 pounds of apples cost?

   a. $8.65
   b. $10.30
   c. $11.55
   d. $11.95

4. If a gallon of gas costs $2.85, how many gallons of gas can Ian buy if he only has $30? (Round your answer to the nearest hundredth.)

   a. 5.26 pounds
   b. 7.38 pounds
   c. 10.53 gallons
   d. 21.45 gallons

5. Alliya pays a monthly membership fee of $8 for an online audio book rental app that charges $3 per audio book. How many books did Alliya rent in one month if her total bill is $20?

   a. 3 books
   b. 4 books
   c. 5 books
   d. 6 books

6. Josh works at a coffee shop downtown and is paid $14 per hour. The coffee shop pays the city $35 a month for each employee to park downtown and then deducts the amount from the employee's paycheck at the end of the month. What is Josh's monthly pay if he works 164 hours in one month?

   a. $2,085
   b. $2,261
   c. $2,296
   d. $2,331

7. Grey works at a boutique where she is paid $17 an hour. The boutique provides parking and lunch each day Grey works, then at the end of the month the boutique deducts $100 from Grey's paycheck. How many hours did Grey work if her monthly pay is $2,960?
   a. 168 hours
   b. 174 hours
   c. 180 hours
   d. 185 hours

8. Zola has a savings account balance of $500. Since she has a new job, she has decided to save $150 a month from her paycheck and deposit it into her savings account. How much money will Zola have in her savings account after 8 months?
   a. $650
   b. $1,700
   c. $4,150
   d. $5,200

9. Nadya is saving up for a trip to Europe. She has $1,200 in her savings account. She saves $200 per month from her paycheck. How many months will it take Nadya to reach $4,000?
   a. 3 months
   b. 4 months
   c. 10 months
   d. 14 months

10. Sarah's phone company charges her a monthly flat fee of $30 and $0.03 per sent text message per month. How much is Sarah's phone bill if she sends 400 text messages in one month?
    a. $12
    b. $18
    c. $42
    d. $48

11. A certain phone company charges a monthly flat fee of $15 plus $0.05 per sent text message per month. How many text messages would someone have to send for their monthly bill to be $55?
    a. 800 text messages
    b. 900 text messages
    c. 1,100 text messages
    d. 1,400 text messages

12. Lora has a savings account balance of $2,400. She wants to save money to buy a new car. She decides to put $300 from each monthly paycheck into the savings account. Which equation can be used to model this situation where x is the number of months that Lora has been saving money and y is the total amount of money in the savings account?
    a. $y = 300x + 2,400$
    b. $y = 2,400x + 300$
    c. $x = 300y + 2,400$
    d. $x = 2,400y + 300$

13. An online music store charges a monthly membership fee of $10 and $0.20 per song that is downloaded. Which equation can be used to model this situation where x is the number of songs downloaded and y is the total monthly bill?

   a. $x = 0.20y + 10$
   b. $y = 0.20x + 10$
   c. $y = 10x + 0.20$
   d. $x = 10y + 0.20$

14. Lex charges $25 to mow a lawn. He spends $80 a week on fuel for his lawn mower. Which equation can be used to model this situation where x is the number of lawns Lex mows in one week and y is the total amount of money that he makes in one week?

   a. $x = 25y - 80$
   b. $x = 80y - 25$
   c. $y = 25x - 80$
   d. $y = 80x - 25$

15. A musician pays $15 a month to sell their music on an app that pays them $0.89 each time one of their songs is downloaded. Which equation can be used to model this situation where x is the number of songs downloaded in one month and y is the total amount of money paid to the musician by the app?

   a. $y = 0.89x - 15$
   b. $y = 15x - 0.89$
   c. $y = 0.89x + 15$
   d. $y = 15x + 0.89$

2.5 Arithmetic Sequences

PRACTICE QUESTIONS

1. Which statement is true about the common difference for an increasing arithmetic sequence?
    a. The common difference is always zero.
    b. The common difference is always negative.
    c. The common difference is always positive.
    d. The common difference can be positive or negative.

2. Which statement is true about the common difference for a decreasing arithmetic sequence?
    a. The common difference is always zero.
    b. The common difference is always negative.
    c. The common difference is always positive.
    d. The common difference can be positive or negative.

3. Here is an arithmetic sequence: 5, 8, 11, 14, ... What is the next term in the sequence?
    a. 16
    b. 17
    c. 18
    d. 19

4. Here is an arithmetic sequence: 43, 37, 31, 25, ... What is the next term in the sequence?
    a. 17
    b. 18
    c. 19
    d. 20

5. Here is an arithmetic sequence: 3, 12, 21, 30, ... What is the 7th term in the sequence?
    a. 57
    b. 58
    c. 59
    d. 60

6. Here is an arithmetic sequence: 62, 55, 48, 41, ... What is the 6th term in the sequence?
    a. 25
    b. 26
    c. 27
    d. 28

7. Here is an arithmetic sequence: 2, 17, 32, ... What are the next three terms in the sequence?
    a. 47, 62, 77
    b. 50, 65, 70
    c. 47, 52, 57
    d. 50, 65, 80

8. Here is an arithmetic sequence: 88, 83, 78, ... What are the next three terms in the sequence?

   a. 73, 78, 83
   b. 72, 67, 62
   c. 73, 68, 63
   d. 72, 77, 82

9. Here is an arithmetic sequence: 7, 12, 17, 22, ... What is the formula for the sequence?

   a. $a_n = 5 + (n-1)$
   b. $a_n = 7 + (n-1)$
   c. $a_n = 5 + 7(n-1)$
   d. $a_n = 7 + 5(n-1)$

10. Here is an arithmetic sequence: 54, 50, 46, 42, ... What is the formula for the sequence?

    a. $a_n = 4 - 54(n-1)$
    b. $a_n = 54 - 4(n-1)$
    c. $a_n = 4 + 54(n-1)$
    d. $a_n = 54 + 4(n-1)$

11. The formula for an arithmetic sequence is $a_n = 107 - 13(n-1)$. What is the 5th term of the sequence?

    a. 13
    b. 55
    c. 94
    d. 107

12. The formula for an arithmetic sequence is $a_n = 38 + 9(n-1)$. What is the 3rd term of the sequence?

    a. 9
    b. 38
    c. 56
    d. 65

13. The table below shows the relationship between the number of sides of a polygon and the sum of the interior angles of that polygon. What is the formula for the arithmetic sequence that represents this relationship where $n = 1$ when the polygon has 3 sides?

Number of sides	3	4	5
Sum of the interior angles	180°	360°	540°

   a. $a_n = 180(n-1)$
   b. $a_n = 360(n-1)$
   c. $a_n = 180 + 180(n-1)$
   d. $a_n = 180 + 360(n-1)$

14. Michaela takes a new job with a starting salary of $50,000 per year. She receives an annual raise of $2,500. Assuming Michaela receives the same raise each year, what will her annual salary be in six years?

- a. $52,500
- b. $57,500
- c. $60,000
- d. $62,500

15. Sabrina is training for a marathon. She runs 20 minutes on day one, 35 minutes on day two, and 50 minutes on day three. If she continues to increase her minutes at the same rate each day, how many minutes will she run on the 8th day of her training?

- a. 75 minutes
- b. 90 minutes
- c. 105 minutes
- d. 125 minutes

2.6 Geometric Sequences

PRACTICE QUESTIONS

1. Find the sixth term in the following geometric sequence:

$$4, 8, 16, 32, \ldots$$

   a. 64
   b. 256
   c. 164
   d. 128

2. Find the fifth term in the following geometric sequence:

$$2, 6, 18, \ldots$$

   a. 162
   b. 218
   c. 256
   d. 180

3. Find the seventh term in the following geometric sequence:

$$4{,}000, 2{,}000, 1{,}000, 500, \ldots$$

   a. 62.5
   b. 8,000
   c. 40
   d. 125

4. Find the seventh term in the following geometric sequence:

$$2, 10, 50, 250, \ldots$$

   a. 25,625
   b. 16,650
   c. 22,250
   d. 31,250

5. Use the geometric sequence formula to write an equation describing the following sequence:

$$8, 32, 128, \ldots$$

   a. $a_n = 8 \cdot 4^{n-1}$
   b. $a_n = 8 \cdot 8^{n-1}$
   c. $a_n = 4 \cdot 8^{n-1}$
   d. $a_n = 8 \cdot 16^{n-1}$

6. Write an equation to find the twelfth term of the following geometric sequence:

$$9, 36, 144, 576, \ldots$$

   a. $a_{12} = 9 \cdot 4^{12-1}$
   b. $a_{12} = 4 \cdot 12^9$
   c. $a_{12} = 9 \cdot 6^{12-1}$
   d. $a_{12} = 9 \cdot 36^{12-1}$

7. Write an equation to find the 20th term of the following geometric sequence:

$$100, 50, 25, \ldots$$

   a. $a_{20} = 100 \cdot (2)^{20-1}$
   b. $a_{20} = 100 \cdot \left(\frac{1}{2}\right)^{20-1}$
   c. $a_{20} = 100 \cdot (50)^{20-1}$
   d. $a_{20} = 50 \cdot (2)^{20-1}$

8. Write an equation to find the 100th term of the following geometric sequence:

$$\frac{1}{4}, \frac{1}{2}, 1, \ldots$$

   a. $a_{100} = \frac{1}{4} \cdot \left(\frac{1}{2}\right)^{100-1}$
   b. $a_{100} = \frac{1}{4} \cdot 2^{100-1}$
   c. $a_{100} = \frac{1}{2} \cdot 2^{100-1}$
   d. $a_{100} = \frac{1}{4} \cdot \left(\frac{1}{4}\right)^{100-1}$

9. Find the fifth term in the following geometric sequence:

$$3, 6, 12, \ldots$$

   a. $a_5 = 48$
   b. $a_5 = 36$
   c. $a_5 = 16$
   d. $a_5 = 54$

10. Find the fourth term in the following geometric sequence:

$$1, 7, \ldots$$

   a. $a_4 = 350$
   b. $a_4 = 149$
   c. $a_4 = 288$
   d. $a_4 = 343$

11. Find the sixth term in the following geometric sequence:

$$10, 20, 40, \ldots$$

   a. $a_6 = 800$
   b. $a_6 = 3{,}200$
   c. $a_6 = 320$
   d. $a_6 = 160$

12. Find the eighth term in the following geometric sequence:

$$5, 15, \ldots$$

   a. $a_8 = 10{,}935$
   b. $a_8 = 2{,}560$
   c. $a_8 = 12{,}550$
   d. $a_8 = 8{,}515$

13. Kevin is working on a school fundraiser and invites 10 people to attend. Each of those ten people invite five more people to attend. Then, each of *those* people invite five additional people to attend. If this pattern continues, how many people will be invited in the fourth "wave" of invitations? (Kevin's original ten invites is the first "wave".)

   a. $a_4 = 5,000$
   b. $a_4 = 3,500$
   c. $a_4 = 1,250$
   d. $a_4 = 2,250$

14. While eating an apple, Quinn notices that there are five seeds in it. From these seeds, she considers planting five apple trees, which could produce a total of 50 apples with five seeds each. If she planted all of *those* seeds, how many seeds would she have from that generation of trees?

   a. 12,500 seeds
   b. 2,500 seeds
   c. 10,000 seeds
   d. 850 seeds

15. In the medical science laboratory, Alan is studying a type of bacteria which can double in number each day. On Monday, there were 150 bacteria cells. How many cells should there be on Friday?

   a. 15,500 cells
   b. 2,400 cells
   c. 3,000 cells
   d. 2,250 cells

Chapter 3: Linear Equations

3.1 Slope and Rate of Change

PRACTICE QUESTIONS

1. The graph of a linear function passes through the points $(7, 1)$ and $(-2, 3)$. What is the slope of this line?

    a. $\frac{9}{2}$
    b. $-\frac{9}{2}$
    c. $\frac{2}{9}$
    d. $-\frac{2}{9}$

2. The table below represents some points on the graph of a linear function.

x	y
0	4
2	12
4	20
6	28
8	36

What is the slope of this linear function?

   a. 4
   b. 8
   c. $\frac{1}{4}$
   d. $\frac{1}{2}$

3. What is the slope of a line with the equation $3x + 2y = 5$?

   a. $\frac{2}{3}$
   b. -3
   c. 3
   d. $-\frac{3}{2}$

4. The graph of $y = -1.5$ is shown on the coordinate plane below. What is the slope of this line?

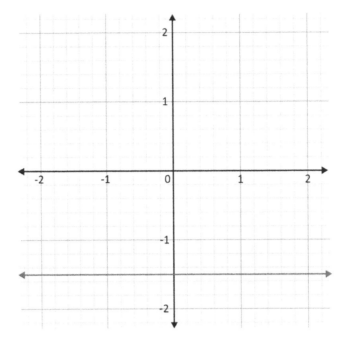

   a. 0
   b. −1.5
   c. 1.5
   d. Undefined

5. What is the slope of a line with the equation $8x + 3y = -9$?

   a. $\frac{8}{3}$
   b. -8
   c. $-\frac{8}{3}$
   d. 8

6. The graph of a linear equation is shown on the coordinate plane below. What is the slope of this line?

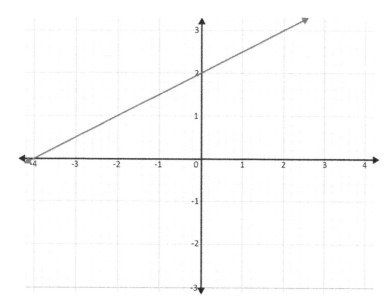

a. 2
b. −2
c. $\frac{1}{2}$
d. $-\frac{1}{2}$

7. The graph of $x = 4$ is shown on the coordinate plane below. What is the slope of this line?

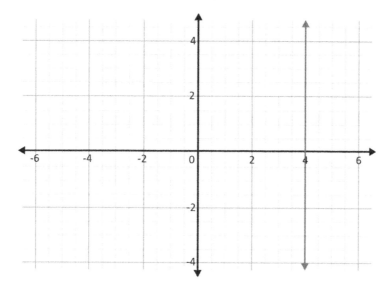

a. 0
b. Undefined
c. 4
d. −4

8. The graph of $y = 2$ is shown on the coordinate plane below. What is the slope of this line?

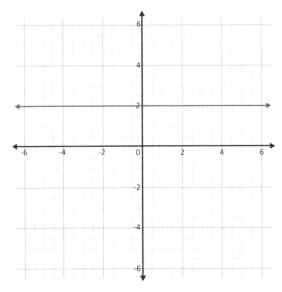

   a. 0
   b. 2
   c. −2
   d. Undefined

9. A taxi driver charges a base fee of $2.70, plus $1.44 for each kilometer traveled. This relationship is represented by the equation $y = 1.44x + 2.70$. Which of the following best describes the rate of change?

   a. The taxi driver charges $4.14 per kilometer.
   b. The taxi driver charges $2.70 per kilometer.
   c. The taxi driver charges $1.44 per kilometer.
   d. The taxi driver charges $1.44 for every 2 kilometers traveled.

10. Maria decides to rent a bicycle for the day. The bicycle rental company charges a flat fee of $8, as well as $1.50 per hour rented. Given this information, determine the direction in which the line would slope if this relationship were graphed on a coordinate plane.

   a. Downward
   b. Upward
   c. Zero
   d. Undefined

11. The Chatterley family went on a road trip. The table below shows the relationship between the amount of time they spent driving in hours (x) and the distance traveled in miles (y).

Time Driving in Hours (x)	Distance Traveled in Miles (y)
0	0
2	40
4	80
6	120

What is the rate of change of the distance traveled per hour?
   a. 120 miles per hour
   b. 80 miles per hour
   c. 40 miles per hour
   d. 20 miles per hour

12. A savings account balance is modeled by the linear function shown on the graph below.

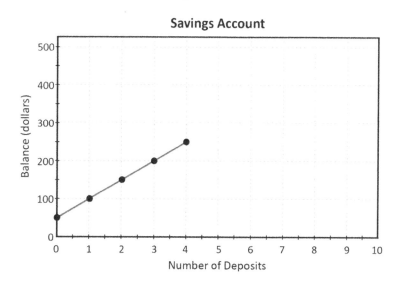

What is the rate of change of the balance per deposit?
   a. $5 per deposit
   b. $0.50 per deposit
   c. $100 per deposit
   d. $50 per deposit

13. Grace runs around a 100-meter track. She runs one warm-up lap, then proceeds with her workout. She uses the linear equation $y = 100 + 4.5x$ to describe the distance, y, she runs x seconds after her warm-up lap. Which of the following best describes the rate of change of the linear equation?
   a. Grace runs 4.5 meters in 2 seconds.
   b. Grace runs 4.5 meters in 1 second.
   c. Grace runs 9 meters in 1 second.
   d. Grace runs 100 meters in 1 second.

14. The table below represents some points on the graph of a linear function.

x	y
1	−14
3	−6
5	2
7	10

What is the rate of change of this function?

   a. −4
   b. 2
   c. 4
   d. 8

15. Shannon's employer pays an hourly rate plus a monthly stipend of 100 dollars. The equation $y = 100 + 30x$ can be used to determine the total pay, y, that Shannon would be paid for x hours of work in a month. What is the rate of change of Shannon's total pay with respect to the number of hours worked?

   a. 30 dollars per hour
   b. 60 dollars per hour
   c. 70 dollars per hour
   d. 100 dollars per hour

3.2 Graphing Linear Equations

PRACTICE QUESTIONS

1. Which graph represents the linear equation $y = 2x - 3$?

a.
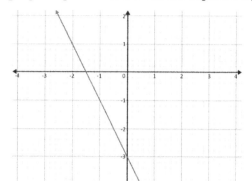

c.
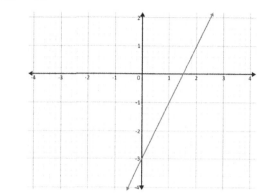

b.
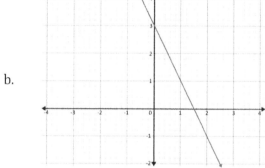

d.
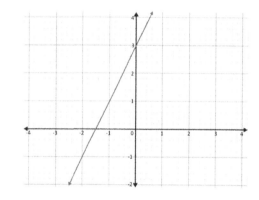

2. The graph of a linear equation is shown on the coordinate plane below.

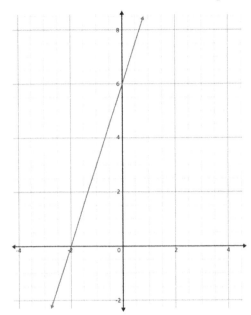

Which ordered pair best represents the location of the x-intercept?

   a. $(6, 0)$
   b. $(-2, 0)$
   c. $(0, -2)$
   d. $(0, 6)$

3. The graph of a linear equation is shown on the coordinate plane below.

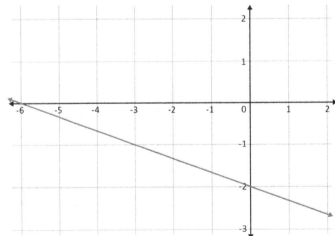

Which ordered pair best represents the location of the y-intercept?

   a. $(-2, 0)$
   b. $(0, -2)$
   c. $(-6, 0)$
   d. $(0, -6)$

4. The graph of a linear equation is shown on the coordinate plane below.

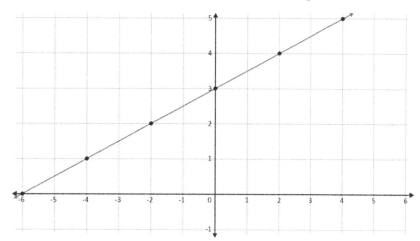

What is the slope of this line?

a. $m = \frac{1}{2}$
b. $m = 2$
c. $m = -\frac{1}{2}$
d. $m = -2$

5. Which graph represents the linear equation $x = 4$?

a.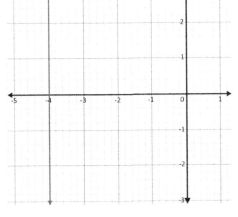

b.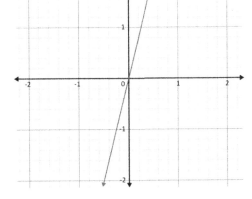

c.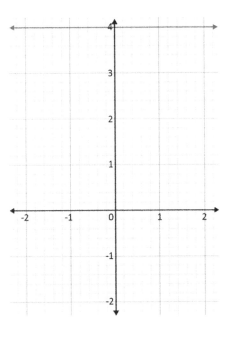

d.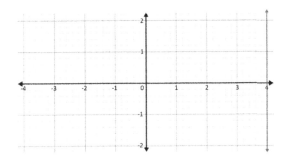

6. Which graph represents the linear equation $2x + 4y = 8$?

a.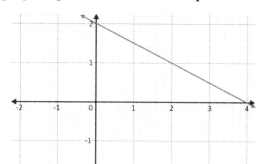

c.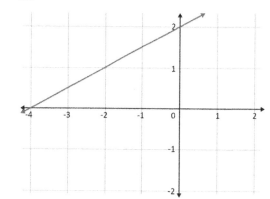

b.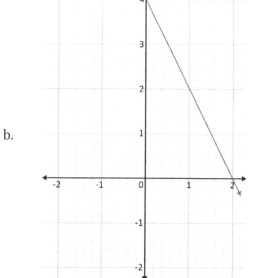

d.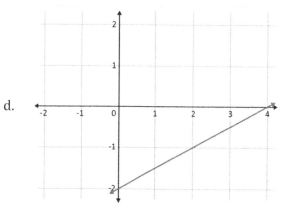

7. The graph of a linear function is shown below.

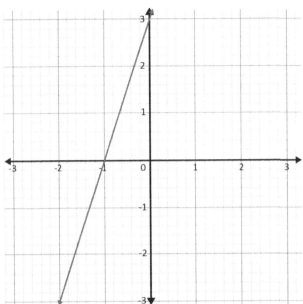

Which coordinate pair best represents the location of the zero of this linear function?

a. $(0, -1)$
b. $(3, 0)$
c. $(-1, 0)$
d. $(0, 3)$

8. Which graph represents the linear equation $y = -6$?

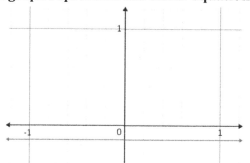
a.

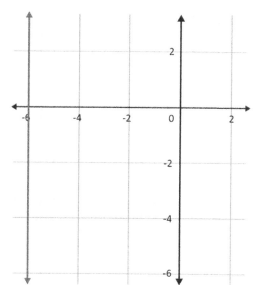
c.

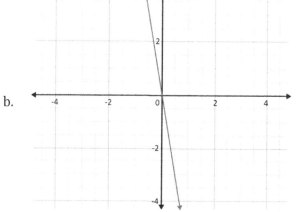
b.

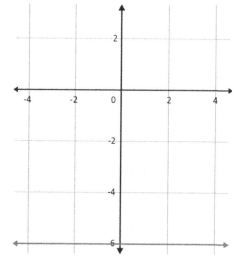
d.

9. What is the y-intercept of the linear equation $4x + 2y = 6$?
   a. $(0, 4)$
   b. $(0, 2)$
   c. $(0, 3)$
   d. $\left(0, 1\frac{1}{2}\right)$

10. The graph of the linear equation $3x + 4y = 1$ is shown on the coordinate plane below.

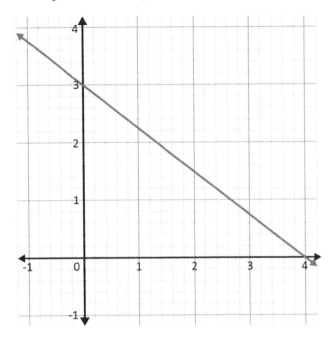

Which ordered pair best represents the location of the y-intercept?
   a. $(3, 0)$
   b. $(0, 3)$
   c. $(4, 0)$
   d. $(0, 4)$

11. Which graph represents the linear equation $-6x + 3y = 15$?

a.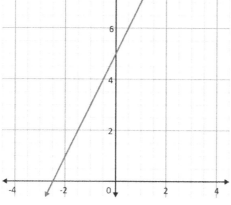

c.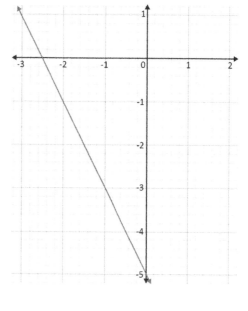

b.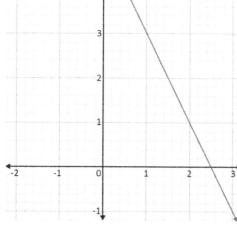

d.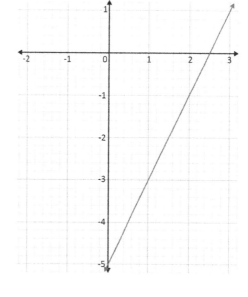

12. What is the x-intercept of the linear equation $2x + 4y = 12$?
   a. $(6, 0)$
   b. $(4, 0)$
   c. $(3, 0)$
   d. $(2, 0)$

13. Mike and Christina are graphing linear equations and identifying their slopes. After graphing the equation $y = 3.5$, they disagree about the slope of the line. Mike thinks the slope is 0, and Christina thinks the slope is undefined. Who is correct?
   a. Christina is correct because a vertical line always has an undefined slope.
   b. Christina is correct because a slope with a decimal is always undefined.
   c. Mike is correct because a vertical line always has a slope of 0.
   d. Mike is correct because a horizontal line always has a slope of 0.

14. The graph below shows how Matthew's savings account balance has changed over the course of a year. He opened his savings account with $250, and after 12 months, he had $2,650 in his savings account. How much money does Matthew save each month?

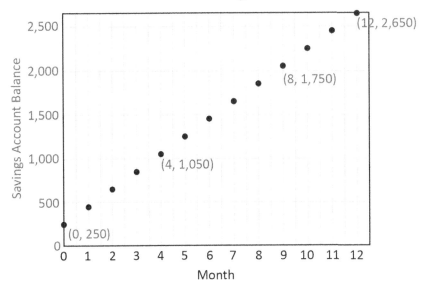

a. Matthew spent $200 each month.
b. Matthew saved $1,200 each month.
c. Matthew saved $200 each month.
d. Matthew saved $250 each month.

15. Marissa goes for a jog with her dog every morning. The distance jogged can be modeled by the equation $d = \frac{1}{10}t$, where d is the distance jogged in miles, and t is the time jogged in minutes. Which graph best represents the relationship between minutes and distance?

a.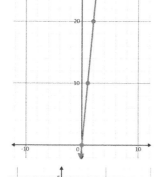

b.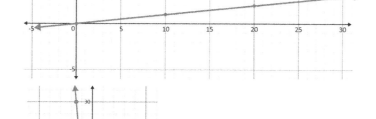

c.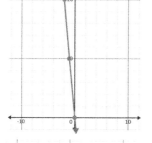

d.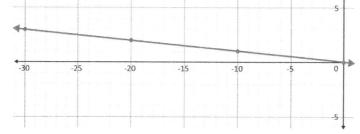

3.3 Finding the Slope-Intercept Form Equation of a Line

PRACTICE QUESTIONS

1. Write the equation for the line that has a slope of –2 and passes through the point $(6, -3)$.
    a. $y = -2x + 9$
    b. $y = 2x$
    c. $y = -x + 2$
    d. $y = 2x + 9$

2. Write the equation for the line that has a slope of 2 and passes through the point $(-1, -6)$.
    a. $y = 2x - 4$
    b. $y = \frac{1}{2}x - 4$
    c. $y = 2x + 4$
    d. $y = 2x - \frac{1}{4}$

3. Write the equation for the line that has a slope of 2 and passes through the point $(-3, 5)$.
    a. $y = 2x - 1$
    b. $y = 2x + 11$
    c. $y = -2x + 1$
    d. $y = 2x + \frac{1}{11}$

4. Write the equation for the line that has a slope of –1 and passes through the point $(-2, 5)$.
    a. $y = -x + 3$
    b. $y = -\frac{1}{2}x + 3$
    c. $y = -x - 3$
    d. $y = -x + 2$

5. Write the equation for the line that has a slope of –3 and passes through the point $(1, 1)$.
    a. $y = -3x$
    b. $y = -3x - 4$
    c. $y = 3x + 4$
    d. $y = -3x + 4$

6. Write the equation for the line that has a slope of $-\frac{5}{3}$ and passes through the point $(4, -1)$.
    a. $y = 3x + 17$
    b. $y = \frac{5}{3}x - \frac{17}{3}$
    c. $y = -\frac{5}{3}x + \frac{17}{3}$
    d. $y = -5x - \frac{17}{3}$

7. Write an equation for the line that passes through the two points $(-4, -2)$ and $(1, 3)$.
    a. $y = 2x + 2$
    b. $y = x - 2$
    c. $y = x + 2$
    d. $x = y + 2$

8. Write an equation for the line that passes through the two points $(-3, 3)$ and $(3, -1)$.
   a. $y = -\frac{2}{3}x + \frac{1}{3}$
   b. $y = -\frac{2}{3}x - 1$
   c. $y = \frac{2}{3}x + 1$
   d. $y = -\frac{2}{3}x + 1$

9. Write an equation for the line that passes through the two points $(1, 6)$ and $(3, 2)$.
   a. $y = \frac{1}{2}x + 8$
   b. $y = -2x + 8$
   c. $y = -2x + 0.5$
   d. $y = 0.2x + 8$

10. Write an equation for the line that passes through the two points $(-5, 2)$ and $(-1, 0)$.
    a. $y = -\frac{1}{2}x - \frac{1}{2}$
    b. $y = -x - \frac{1}{2}$
    c. $y = -\frac{1}{2}x - 1$
    d. $x = -\frac{1}{2}y - \frac{1}{2}$

11. Write an equation for the line that passes through the two points $(3, 2)$ and $(9, 7)$.
    a. $y = \frac{1}{2}x - \frac{1}{2}$
    b. $y = \frac{5}{6}x + \frac{1}{2}$
    c. $y = \frac{5}{6}x - \frac{1}{2}$
    d. $y = -\frac{5}{6}x - 2$

12. Write an equation for a line that passes through the two points $(1, -7)$ and $(3, -15)$.
    a. $y = 0.4x - 3$
    b. $y = -4x + 1$
    c. $y = -\frac{4x}{3}$
    d. $y = -4x - 3$

13. Evan is emptying out a fish tank. The tank starts out with 175 gallons of water and gradually empties at a steady rate. The graph below shows gallons of water on the y-axis and time (in hours) on the x-axis. The point $(3, 100)$ represents 100 gallons of water after 3 hours. The point $(7, 0)$ represents 0 gallons of water after 7 hours. Which equation below represents the rate of water emptying?

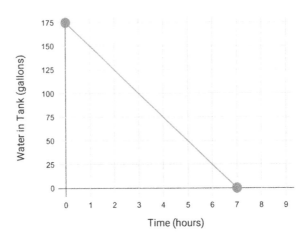

a. $y = 25x - 175$
b. $y = -25x + 175$
c. $y = 25x + 175$
d. $y = -25x + 7$

14. Gloria is raising a flag to the top of the flagpole in front of her school. The flag starts at a height of 3 feet and gradually raises as she pulls the ropes attached to the pole. The point $(2, 6)$ represents the flag height of 6 feet after 2 seconds. Write an equation in slope-intercept form based on the scenario.

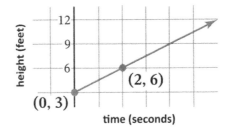

a. $y = \frac{2}{3}x + 3$
b. $y = \frac{3}{2}x + 3$
c. $y = 2x + 3$
d. $y = \frac{3}{2}x + 2$

15. Carl charges $45 per hour as a plumber. On a graph, the point $(0, 100)$ represents the one-time fee of $100 that Carl charges before any hours of work have been completed. Write an equation in slope-intercept form where x is the hours Carl works on a job and y is the total amount Carl charges for the job.

   a. $y = 45x + 100$
   b. $y = 45x$
   c. $y = 4.5x + 100$
   d. $y = 45x - 100$

3.4 Finding the Point-Slope Form Equation of a Line

PRACTICE QUESTIONS

1. Write an equation in point-slope form for a line that has a slope of $\frac{1}{2}$ and passes through the point $(-3, 6)$.

   a. $y = \frac{1}{2}(x + 3)$
   b. $y + 6 = \frac{1}{2}(x - 3)$
   c. $y - 3 = \frac{1}{2}(x - 3)$
   d. $y - 6 = \frac{1}{2}(x + 3)$

2. Write an equation in point-slope form for a line that has a slope of 2 and passes through the point $(-3, 1)$.

   a. $y - 1 = 2(x - 3)$
   b. $y + 1 = 2(x + 3)$
   c. $x - 1 = 2(y + 3)$
   d. $y - 1 = 2(x + 3)$

3. Write an equation in point-slope form for a line that has a slope of $\frac{3}{2}$ and passes through the point $(0, 2)$.

   a. $y - 2 = \frac{3}{2}(x - 0)$
   b. $y + 2 = \frac{3}{2}(x - 1)$
   c. $x - 2 = \frac{3}{2}(y - 0)$
   d. $y - \frac{3}{2} = 2(x - 0)$

4. When writing an equation in point-slope form, $y - y_1 = m(x - x_1)$, the variable m represents the _____, and the variables x_1 and x_2 represent _____.

   a. slope/y-intercept
   b. y-intercept/a point on the line
   c. slope/a point on the line
   d. an ordered pair/another point on the line

5. A line goes through the point $(1, 4)$. If the slope of the same line is 5, what is the equation in point-slope form?

   a. $y - 4 = 5(x - 1)$
   b. $y = 5x + 4$
   c. $y + 4 = 5(x + 1)$
   d. $-4x + 5x = -1$

6. Write an equation in point-slope form for a line that has a slope of 2 and passes through the point $\left(\frac{1}{2}, 1\right)$.
   a. $y - 1 = 2(x - 1)$
   b. $y - 1 = 2\left(x - \frac{1}{2}\right)$
   c. $y + 1 = \frac{1}{2}(x - 2)$
   d. $y = 2\left(x - \frac{1}{2}\right)$

7. Which of the following is an equation of a line in point-slope form that passes through the points $(0, -2)$ and $(3, 4)$?
   a. $y = 2(x - 3)$
   b. $y - 4 = 4(x + 3)$
   c. $y + 4 = (x - 3)$
   d. $y - 4 = 2(x - 3)$

8. Which of the following is an equation in point-slope form that passes through the points $(-3, 5)$ and $(2, 8)$?
   a. $y + 8 = \frac{3}{5}(x + 2)$
   b. $y = \frac{3}{5}x - 2$
   c. $y - 2 = \frac{3}{5}\left(x - \frac{3}{5}\right)$
   d. $y - 8 = \frac{3}{5}(x - 2)$

9. Which of the following is an equation of a line in point-slope form that passes through the points $(-2, -2)$ and $(4, 1)$?
   a. $y - 1 = x - 4$
   b. $y - 1 = \frac{1}{2}(x + 4)$
   c. $y = \frac{1}{2}(x - 4)$
   d. $y - 1 = \frac{1}{2}(x - 4)$

10. What is the correct way to write an equation in point-slope form for the two points $(-2, -3)$ and $(4, -2)$?
    a. $y + 2 = \frac{1}{6}(x - 2)$
    b. $y + 3 = \frac{1}{6}(x + 2)$ or $y + 2 = 2(x - 2)$
    c. $y + 2 = \frac{1}{6}(x - 4)$ or $y + 3 = \frac{1}{6}(x + 2)$
    d. $y = \frac{1}{6}x + 2$

11. What is an equation of a line in point-slope form that passes through the points $(3, 3)$ and $(-3, -3)$.
    a. $y - 3 = (x - 3)$
    b. $y = x - 3$
    c. $3y - 3x = 1$
    d. $y - 3 = (x + 3)$

12. What is an equation of a line in point-slope form that passes through the points $(4, -7)$ and $(1, 2)$.
    a. $y = -3(x - 1)$
    b. $y - 2 = 3(x - 1)$
    c. $y - 2 = -3\left(x - \frac{1}{3}\right)$
    d. $y - 2 = -3(x - 1)$

13. Eliza is making necklaces to sell at her community market. Eliza wants to sell larger orders of necklaces, so her new sales strategy is to sell two necklaces for $10 and then $3 for every necklace after that. She hopes this will encourage customers to buy multiple necklaces. Express the scenario in point-slope form where x is the number of necklaces in each order and y is the cost of the order.
    a. $y + 10 = 3(x + 2)$
    b. $y - 10 = 3(x - 2)$
    c. $y - 2 = 3(x - 10)$
    d. $y - 1 = 10(x - 2)$

14. Gregory is on a work trip, and the company that he works for allows him to use a pre-loaded company debit card to spend $25 per day for his lunch expenses. On the 4th day of his trip, he has $175 left on the debit card. Express the scenario in point-slope form where x is the day of the trip and y is the amount of money on the debit card.
    a. $y = -25x + 175$
    b. $y = -25x - 4$
    c. $y - 4 = -25(x - 4)$
    d. $y - 175 = -25(x - 4)$

15. Jamie and Julia disagree on the correct way to write an equation in point-slope form if the slope is 3 and a point on the line is $(-2, 5)$. Jamie is confident that the equation would be $y - 5 = 3(x - 2)$. However, Julia states that the equation is $y - 5 = 3(x + 2)$. Who is correct?
    a. Jamie and Julia are both correct.
    b. Jamie and Julia are both incorrect.
    c. Jamie is correct: $y - 5 = 3(x - 2)$.
    d. Julia is correct: $y - 5 = 3(x + 2)$.

3.5 Finding the Standard Form Equation of a Line

Practice Questions

1. Write an equation in standard form that has a slope of -3 and passes through the point $(1, 5)$.

   a. $9x + y = 8$
   b. $3x + y = 9$
   c. $x + 3y = 8$
   d. $3x + y = 8$

2. Write an equation in standard form that has a slope of $-\frac{1}{3}$ and passes through the point $(4, -1)$.

   a. $3x - y = \frac{1}{3}$
   b. $x + 3y = 1$
   c. $x + y = 1$
   d. $\frac{1}{3}x - y = \frac{1}{3}$

3. Write an equation in standard form that has a slope of 2 and passes through the point $(3, 5)$.

   a. $\frac{1}{2}x - y = 2$
   b. $2x + y = -1$
   c. $-3x - y = 2$
   d. $2x - y = 1$

4. Write an equation in standard form that has a slope of 5 and passes through the point $(6, -9)$.

   a. $5x - y = 39$
   b. $-5x + y = 39$
   c. $5x - y = 24$
   d. $-5x + y = 24$

5. Write an equation in standard form that has a slope of $-\frac{3}{7}$ and passes through the point $(1, -2)$.

   a. $\frac{1}{3}x + \frac{1}{7}y = -11$
   b. $3x - 7y = 11$
   c. $3x + 7y = -11$
   d. $3x + 7y = -\frac{1}{3}$

6. The example below shows a conversion from point-slope form to standard form. Identify the error.

   Point: (2,3) and Slope = 4
   Step 1: $y - 3 = 4(x - 2)$
   Step 2: $y - 3 = 4x - 8$
   Step 3: $-4x + y = -5$

   a. The constant, –5, needs to be on the left side of the equation in step 3.
   b. The constant cannot be negative in step 3.
   c. A cannot be negative in step 3.
   d. There is no error.

7. Write an equation in standard form that passes through the two points $(2, 1)$ and $(-1, 3)$.
   a. $2x - 3y = 9$
   b. $2x + 3y = 7$
   c. $3x + 2y = -7$
   d. $y = 3x + 2y$

8. Write an equation in standard form that passes through the two points $(-1, 3)$ and $(4, -2)$.
   a. $2x + 4y = -2$
   b. $x + 3y = -3$
   c. $x + y = 2$
   d. $2x + y = 2$

9. Write an equation in standard form that passes through the two points $(-1, -5)$ and $(-4, 1)$.
   a. $2x + 3y = \frac{1}{7}$
   b. $-2x + y = 7$
   c. $-7x + y = 3$
   d. $2x + y = -7$

10. Write an equation in standard form that passes through the two points $(-1, 5)$ and $(0, 8)$.
    a. $3x - y = -8$
    b. $\frac{1}{3}x - y = -8$
    c. $3x + y = 8$
    d. $x - y = 8$

11. Write an equation in standard form that passes through the two points $(1, 3)$ and $(3, 7)$.
    a. $5x - y = -1$
    b. $4x - y = -1$
    c. $3x - y = -1$
    d. $2x - y = -1$

12. Write an equation in standard form that passes through the two points $(-2, 3)$ and $(-4, -5)$.
    a. $3x - y = 11$
    b. $4x + 3y = -11$
    c. $4x - y = -11$
    d. $2x - y = 11$

13. Your school needs to raise $5,000 for a new soccer stadium. A fundraiser is selling $200 student tickets and $300 adult tickets for a concert to raise money for the new stadium. Write an equation in standard form that could be used to determine the number of student and adult tickets that need to be sold to reach the $5,000 goal exactly.

   a. $20.0x + 30.0y = 5,000$
   b. $200x - 300y = 5,000$
   c. $200x + 300y = 5,000$
   d. $250x + 350y = 5,000$

14. Jamie sells snacks during half-time at a basketball game. He sells pizza slices for $1 and bags of popcorn for 50¢. After the game, he has earned a total of $200. Write an equation in standard form that can be used to determine the number of pizza slices, x, and popcorn bags, y, he sold.

   a. $2x + 0.5y = 200$
   b. $2x + y = 400$
   c. $y + 0.5x = 2.00$
   d. $x + 5y = 400$

15. The following equation is in standard form: $2x + 5y = 50$. Match the equation with one of the written descriptions below.

   a. Robin spent $2 on ice cream and $5 on fries. Robin has $50 left over.
   b. April earned $5 per hour pulling weeds out of the garden. She worked for 50 hours and took a 2-hour break.
   c. Susan spent $50 on flowers at the nursery. Each outdoor plant cost $2 and each indoor plant cost $6.
   d. Dave has $50 dollars to spend on lunch. Drinks cost $2 each and sandwiches cost $5 each.

3.6 Converting Between Standard Form and Slope-Intercept Form

PRACTICE QUESTIONS

1. Convert the following equation from standard form to slope-intercept form: $6x + 2y = 4$.
   a. $y = 3x - 2$
   b. $y = -3x + 2$
   c. $y = -\frac{1}{3}x + 2$
   d. $y = -2x + \frac{1}{2}$

2. Convert the following equation from standard form to slope-intercept form: $3x - 2y = 21$.
   a. $y = 4x - \frac{21}{2}$
   b. $y = \frac{3}{2}x - \frac{21}{2}$
   c. $y = \frac{3}{2}x - 21y$
   d. $y = \frac{21}{2}x - 4$

3. Convert the following equation from standard form to slope-intercept form: $8x + 4y = 16$.
   a. $y = 2x + 4$
   b. $y = -2x + 4$
   c. $y = -2x - 4$
   d. $y = -4x + 2$

4. Convert the following equation from standard form to slope-intercept form: $x + y = 7$.
   a. $y = x + 7$
   b. $y = x$
   c. $y = -x + 7$
   d. $y = x - 7$

5. Convert the following equation from standard form to slope-intercept form: $10x - 7y = -35$.
   a. $y = \frac{10}{7}x + 5$
   b. $y = x + 5$
   c. $y = \frac{10}{7}x$
   d. $y = 1.7x + 5$

6. Convert the following equation from standard form to slope-intercept form: $2x - 3y = -6$.
   a. $x = \frac{2}{3}y + 2$
   b. $y = \frac{2}{3}x + 2$
   c. $y = \frac{2}{3}x$
   d. $y = \frac{2}{3}x + \frac{1}{2}$

7. When converting from standard form to slope-intercept form, which variable should you isolate using inverse operations?
   a. x
   b. y
   c. A
   d. B

8. Convert the equation $y = 5x - 9$ from slope-intercept form to standard form.
   a. $-9x + y = -5$
   b. $9x - y = 5$
   c. $5x - y = 9$
   d. $-5x + y = -9$

9. Below is a step-by-step process for converting from standard form to slope-intercept form. Identify the step that has an error.
   Step 1: $6x + 3y = 1$
   Step 2: $3y = -6x + 1$
   Step 3: $y = -2x + 1$
   a. Error in Step 1
   b. Error in Step 2
   c. Error in Step 3
   d. There is no error.

10. Convert the following equation from slope-intercept form to standard form: $y = \frac{2}{3}x + 3$.
    a. $3x - 2y = -9$
    b. $-\frac{2}{3}x + y = 3$
    c. $-3y + 2x = 9$
    d. $2x - 3y = -9$

11. Convert the following equation from slope-intercept form to standard form: $y = -\frac{3}{8}x + 6$.
    a. $3x + 8y = 48$
    b. $6x + 3y = -8$
    c. $3x - 8y = -48$
    d. $\frac{3}{8}x + y = 6$

12. Convert the following equation from slope-intercept form to standard form: $y = \frac{12}{5}x + \frac{44}{5}$.
    a. $12x - 5y = -44$
    b. $-y + 12x = 44$
    c. $-\frac{12}{5}x + y = \frac{44}{5}$
    d. $-12x + y = 44$

13. Ryan is converting from slope-intercept form to standard form. His original equation is $y = 6x - 3$. Ryan starts by subtracting $6x$ from both sides of the equation, which results in $-6x + y = -3$. What is Ryan's next step?
   a. Add 3 to both sides.
   b. Divide both sides by –6.
   c. Subtract y from both sides.
   d. Multiply both sides by –1.

14. Lauren and Jose disagree on whether standard form or slope-intercept form is more convenient for graphing. Lauren says that it is best to convert equations from slope-intercept to standard form and then graph. Jose says that it is best to convert equations from standard form to slope-intercept form and then graph. Who is correct?
   a. Lauren is correct.
   b. Jose is correct.
   c. Neither Lauren nor Jose is correct.
   d. Lauren and Jose are both correct.

15. Marci is at a bookstore buying books and bookmarks for her family members. The books she purchases cost $20 each and the bookmarks cost $1 each. Marci spends $80 at the bookstore. The equation that represents this scenario is $20x + y = 80$. Convert the equation to slope-intercept form.
   a. $y = \frac{1}{80}x + \frac{1}{4}$
   b. $x = -\frac{1}{20}y + 4$
   c. $y = -20x + 80$
   d. $y = -80x + 20$

3.7 Writing Linear Equations

PRACTICE QUESTIONS

1. Write a linear equation based on the table of values.

x	y
−2	−4
−1	−1
0	2
1	5
2	8

   a. $y = 3x + 2$
   b. $y = 2x + 3$
   c. $y = \frac{1}{3}x - 2$
   d. $y = -3x - 2$

2. Write a linear equation based on the table of values.

x	y
1	6
2	12
3	18

   a. $y = 6x$
   b. $y = 0.6x$
   c. $y = \frac{1}{6}x$
   d. $x = 6y$

3. Use the table below to write a linear equation.

x	y
0	−7
1	−5
2	−3
3	−1
4	1

   a. $y = -2x - 7$
   b. $y = \frac{1}{2}x - 7$
   c. $y = 7x - 2$
   d. $y = 2x - 7$

4. Write a linear equation that represents the data table below.

x	y
1	11
2	15
3	19
4	23
5	27

   a. $y = 4x$
   b. $y = 2x + 7$
   c. $y = 4x + 7$
   d. $y = \frac{1}{4}x + 7$

5. Write a linear equation that describes the graph.

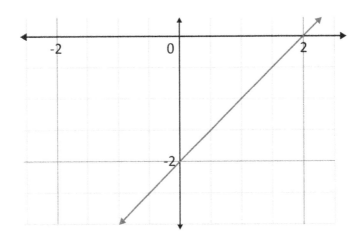

   a. $y = x + 2$
   b. $y = x - 2$
   c. $x = y - 2$
   d. $y = 2x - 2$

6. Write a linear equation based on the graph.

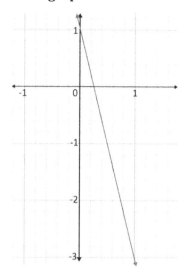

a. $y = 5x + \frac{1}{4}$
b. $y = 4x - 1$
c. $y = -4x + 1$
d. $y = \frac{1}{4}x + 3$

7. Write a linear equation that describes the graph.

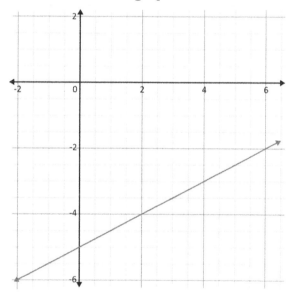

a. $y = 2x - 5$
b. $y = \frac{1}{2}x + 5$
c. $y = \frac{1}{2}x - 5$
d. $y = -\frac{1}{2}x + 5$

8. Write a linear equation that describes the graph.

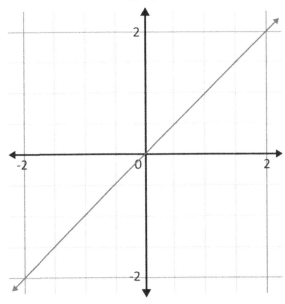

   a. $y = x$
   b. $y = \frac{1}{2}x$
   c. $y = 1x + 1$
   d. $y = x + 1$

9. **Write a linear equation that represents the following description:**

   Damion is arranging a sailboat tour for his vacation. The sailboat company charges a one-time fee of $150 to reserve the boat and $25 per person for the tour. Damion has 6 family members joining him on the vacation. Write a linear equation to calculate the total cost of the sailing tour for Damion and his family.

   a. $x = 25y + 150$
   b. $y = 25x - 150$
   c. $y = 5x + 150$
   d. $y = 25x + 150$

10. **Michael is hiking up a mountain at a steady rate of 3 miles per hour. How many miles has Michael traveled if he hikes for 2.5 hours?**

    a. 6.25 miles
    b. 7.5 miles
    c. 8.5 miles
    d. 0.75 miles

11. **Margaret can bake 42 cookies in 2 hours. If she bakes cookies for x hours and then eats 3 of the cookies, how many cookies remain (y)?**

    a. $y = 21x + 3$
    b. $y = 21x - 3$
    c. $y = 42x - 3$
    d. $y = 42x + 3$

12. Edgar owns a car wash that charges $24 per wash. Today, one of Edgar's most valued customers was so pleased with the quality of service that he left a $50 tip! How much has Edgar's car washing business earned after 20 washes and the tip?
    a. $150
    b. $265
    c. $530
    d. $1,024

13. A parking garage charges a flat rate of $2 and then $0.50 for every hour a car is parked there. If Andrea parks her car in the garage at 2:00 p.m. and then leaves the garage at 7:00 p.m., how much is her bill?
    a. $4.50
    b. $5.00
    c. $4.50
    d. $6.50

14. Susan has been recording the number of t-shirts she sells each week. She currently sells 8 shirts per week. As she continues to grow her small business, her goal is to sell at least 50 t-shirts after 7 weeks. Write a linear equation based on the graph and determine if Susan will meet her goal.

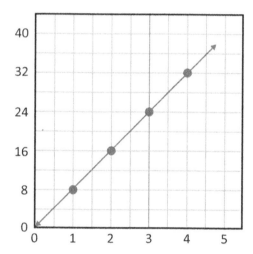

   a. $y = 8x + 7$; no, Susan will not meet her goal.
   b. $y = 8x + 1$; yes, Susan will meet her goal.
   c. $y = x + 8$; no, Susan will not meet her goal.
   d. $y = 8x$; yes, Susan will meet her goal.

15. Erika owns a yoga studio. The table below shows the number of enrolled members at her studio each year. Erika wants to write a linear equation from the table so that she can predict the number of members she will have after 7 years of business. Using an equation that describes the table, how many members will Erika have after 7 years?

Time (years)	0	1	2	3	4
Number Enrolled	8	11	14	17	20

a. 23
b. 26
c. 29
d. 31

3.8 Parallel and Perpendicular Lines and the Axes

PRACTICE QUESTIONS

1. Which equation of a line is parallel to the x-axis, and what is its slope?
   a. $y = 8; m = 0$
   b. $y = 8x; m = 0$
   c. $x = 8; m = $ undefined
   d. $x = 8y; m = $ undefined

2. Which equation of a line is parallel to the y-axis, and what is its slope?
   a. $y = 4; m = 0$
   b. $y = 4x; m = 0$
   c. $x = 4; m = $ undefined
   d. $x = 4y; m = $ undefined

3. Which equation of a line is perpendicular to the x-axis, and what is its slope?
   a. $y = 2x; m = 0$
   b. $x = 2y; m = 0$
   c. $y = 2; m = $ undefined
   d. $x = 2; m = $ undefined

4. Which equation of a line is perpendicular to the y-axis, and what is its slope?
   a. $x = 1; m = 0$
   b. $x = 1; m = $ undefined
   c. $y = 1; m = 0$
   d. $y = 1; m = $ undefined

5. Which equation of a line is parallel to the y-axis, and what is its slope?
   a. $y = -3x; m = 0$
   b. $y = -3; m = 0$
   c. $x = -3y; m = $ undefined
   d. $x = -3; m = $ undefined

6. Which equation of a line is perpendicular to the x-axis, and what is its slope?
   a. $x = 12; m = $ undefined
   b. $y = 12; m = $ undefined
   c. $x = 12; m = 0$
   d. $y = 12; m = 0$

7. Which equation of a line is parallel to the x-axis, and what is its slope?
   a. $x = 7.5y; m = $ undefined
   b. $x = 7.5; m = $ undefined
   c. $y = 7.5x; m = 0$
   d. $y = 7.5; m = 0$

8. Which equation of a line is perpendicular to the y-axis, and what is its slope?
   a. $y = 6; m = 0$
   b. $y = 6x; m = 0$
   c. $x = 6; m = $ undefined
   d. $x = 6y; m = $ undefined

9. Which equation of a line is perpendicular to the x-axis, and what is its slope?
   a. $y = -1; m = $ undefined
   b. $x = -1; m = $ undefined
   c. $y = -1; m = 0$
   d. $x = -1; m = 0$

10. Which equation of a line is perpendicular to the y-axis, and what is its slope?
    a. $y = -10; m = $ undefined
    b. $y = -10; m = 0$
    c. $x = -10; m = $ undefined
    d. $x = -10; m = 0$

11. Which equation of a line is parallel to the y-axis, and what is its slope?
    a. $y = -5.5; m = $ undefined
    b. $y = -5.5x; m = 0$
    c. $x = -5.5; m = $ undefined
    d. $x = -5.5; m = 0$

12. Which equation of a line is parallel to the x-axis, and what is its slope?
    a. $y = -11x; m = $ undefined
    b. $y = -11; m = 0$
    c. $x = -11y; m = $ undefined
    d. $x = -11; m = 0$

13. A hiker is using a coordinate plane to analyze a map of a state park. The map shows several different hiking trails within the park. The green hiking trail runs along the x-axis, and the red hiking trail runs perpendicular to the green hiking trail. The green and red hiking trails intersect at point $(11.4, 0)$ on the coordinate plane. What is the linear equation of the red hiking trail, and what is its slope?
    a. $x = 11.4; m = $ undefined
    b. $y = 11.4; m = $ undefined
    c. $x = 11.4; m = 0$
    d. $y = 11.4; m = 0$

14. A truck driver is using a coordinate plane to analyze a map of his driving route. The map shows freeways in San Francisco. Nimitz Freeway runs along the y-axis, and MacArthur Freeway runs parallel to Nimitz Freeway. What could be the linear equation for MacArthur Freeway, and what is its slope?
    a. $y = 2.8x; m = 0$
    b. $y = 2.8; m = 0$
    c. $x = 2.8y; m = $ undefined
    d. $x = 2.8; m = $ undefined

15. A Pennsylvania Railroad line runs perpendicular to Main Street. When examined on a coordinate plane, the railroad line runs along the x-axis, and Main Street intersects the railroad at point $(-9, 0)$. What is the linear equation for Main Street, and what is its slope?
   a. $x = -9; m = 0$
   b. $y = -9; m = 0$
   c. $x = -9; m =$ undefined
   d. $y = -9; m =$ undefined

3.9 Equations of Parallel Lines

Practice Questions

1. Which equation of a line passes through the point $(3, 2)$ and is parallel to the graph of $y = 6x + 4$?

   a. $y = 6x + 20$
   b. $y = 6x - 16$
   c. $y = -6x - 16$
   d. $y = \frac{1}{6}x - 16$

2. Determine whether the linear equations below are parallel. If so, what is their slope?

$$4x + y = 5$$
$$4x + y = -2$$

   a. The lines are parallel. The slope of both lines is 4.
   b. The lines are parallel. The slope of both lines is –4.
   c. The lines are not parallel because their slopes are different.
   d. The lines are not parallel because they are the same line.

3. Which equation of a line passes through the point $(7, 3)$ and is parallel to the graph of $4x + 2y = 8$?

   a. $y = -4x + 4$
   b. $y = -2x + 11$
   c. $y = -2x + 17$
   d. $y = -4x + 8$

4. Determine whether the linear equations below are parallel. If so, what is their slope?

$$y = \frac{1}{3}x + 4$$
$$3x + y = -4$$

   a. The lines are parallel. The slope of both lines is –3.
   b. The lines are parallel. The slope of both lines is $\frac{1}{3}$.
   c. The lines are not parallel because they are the same line.
   d. The lines are not parallel because their slopes are different.

5. Which equation of a line passes through the point $(1, 1)$ and is parallel to the graph of $3x + 5y = 10$?

   a. $y = 3x + 1\frac{3}{5}$
   b. $y = \frac{5}{3}x + \frac{2}{5}$
   c. $y = -\frac{3}{5}x + 1\frac{3}{5}$
   d. $y = -\frac{3}{5}x + \frac{2}{5}$

6. Which equation of a line passes through point $(4, 5)$ and is parallel to the graph of $8x + 2y = 12$?
   a. $y = -4x + 21$
   b. $y = -4x - 11$
   c. $y = 8x + 21$
   d. $y = -\frac{1}{4}x - 11$

7. Determine whether the linear equations below are parallel. If so, what is their slope?
$$2x + y = 4$$
$$2x + y = \frac{1}{4}$$
   a. These lines are parallel. The slope of both lines is –2.
   b. These lines are parallel. The slope of both lines is 2.
   c. These lines are not parallel because their slopes are different.
   d. The lines are not parallel because they are the same line.

8. Determine whether the linear equations below are parallel. If so, what is their slope?
$$-\frac{1}{4}x - y = -3$$
$$x + 4y = -32$$
   a. The lines are not parallel because they are the same line.
   b. The lines are not parallel because the slopes are different.
   c. The lines are parallel. The slope of both lines is 4.
   d. The lines are parallel. The slope of both lines is $-\frac{1}{4}$.

9. Which equation of a line passes through the point $(-6, 1)$ and is parallel to the graph of $9x + 3y = 1$?
   a. $y = -3x + 19$
   b. $y = -3x - 19$
   c. $y = -3x - 17$
   d. $y = -9x - 53$

10. Determine whether the linear equations below are parallel. If so, what is their slope?
$$\frac{2}{3}x - y = -4$$
$$\frac{3}{2}x + y = 4$$
   a. The lines are parallel. The slope of both lines is $\frac{2}{3}$.
   b. The lines are parallel. The slope of both lines is $-\frac{3}{2}$.
   c. The lines are not parallel because they are the same line.
   d. The lines are not parallel because their slopes are different.

11. Determine whether the linear equations below are parallel. If so, what is their slope?

$$\frac{3}{5}x - y = 2$$
$$3x - 5y = -5$$

   a. The lines are parallel. The slope of both lines is 3.
   b. The lines are parallel. The slope of both lines is $\frac{3}{5}$.
   c. The lines are not parallel because their slopes are different.
   d. The lines are not parallel because they are the same line.

12. Which equation of a line passes through the point $(8, 2)$ and is parallel to the graph of $x + 4y = 12$?

   a. $y = -\frac{1}{4}x$
   b. $y = -\frac{1}{4}x + 4$
   c. $y = 4x + 4$
   d. $y = 4x - 4$

13. When graphed on the coordinate plane, one bike path has the equation $4x - y = 1$, and another bike path has the equation $\frac{1}{4}x - y = 1$. Determine whether these two bike paths are parallel to one another. If so, what is their slope?

   a. The lines are not parallel because their slopes are different.
   b. The lines are not parallel because they are the same line.
   c. The lines are parallel. The slope of both lines is 4.
   d. The lines are parallel. The slope of both lines is $\frac{1}{4}$.

14. When graphed on a coordinate plane, the New Jersey Turnpike has the equation $\frac{1}{2}x + y = -3$. Interstate 295 runs parallel to the New Jersey Turnpike and passes through the point $(12, 3)$. What is the equation of the line for Interstate 295 when graphed on the same coordinate plane?

   a. $y = \frac{1}{2}x + 3$
   b. $y = \frac{1}{2}x - 3$
   c. $y = -\frac{1}{2}x + 9$
   d. $y = -\frac{1}{2}x - 3$

15. When graphed on a coordinate plane, a linear portion of the Schuylkill River has the equation $2x - 6y = -30$. A running trail runs parallel to this portion of the Schuylkill River and passes through the point $(15, 13)$. What is the equation of the line for the running trail?

   a. $y = -3x + 8$
   b. $y = -3x + 18$
   c. $y = \frac{1}{3}x + 8$
   d. $y = \frac{1}{3}x + 18$

3.10 Equations of Perpendicular Lines

PRACTICE QUESTIONS

1. Which equation of a line contains the point $(2, 5)$ and is perpendicular to the graph of $y = 4x + 3$?

    a. $y = -4x + 5\frac{1}{2}$
    b. $y = -4x + 4\frac{1}{2}$
    c. $y = -\frac{1}{4}x + 5\frac{1}{2}$
    d. $y = -\frac{1}{4}x + 4\frac{1}{2}$

2. Which statement best describes the relationship between the linear equations below?
$$\frac{1}{3}x + y = 2$$
$$y = 3x - 8$$

    a. The linear equations are parallel. The slope of both lines is $\frac{1}{3}$.
    b. The linear equations are perpendicular. The slope of both lines is 3.
    c. The linear equations are perpendicular. Their slopes are opposite reciprocals of one another.
    d. The linear equations are neither parallel nor perpendicular.

3. Which equation of a line contains the point $(2, -1)$ and is perpendicular to the graph of $5x + y = 1$?

    a. $y = -\frac{1}{5}x + 1\frac{2}{5}$
    b. $y = \frac{1}{5}x - 1\frac{2}{5}$
    c. $y = -\frac{1}{5}x - \frac{3}{5}$
    d. $y = \frac{1}{5}x - \frac{3}{5}$

4. Which statement best describes the relationship between the linear equations below?
$$7x - y = -2$$
$$x + 7y = 21$$

    a. The linear equations are parallel. The slope of both lines is 7.
    b. The linear equations are neither parallel nor perpendicular.
    c. The linear equations are perpendicular. Their slopes are opposite reciprocals of one another.
    d. The linear equations are perpendicular. The slope of both lines is $-\frac{1}{7}$.

5. Which equation of a line contains the point $(-4, -7)$ and is perpendicular to the graph of $4x + 7y = 3$?

    a. $y = -\frac{1}{4}x - 7$
    b. $y = -\frac{1}{4}x + 14$
    c. $y = \frac{7}{4}x - 14$
    d. $y = \frac{7}{4}x$

6. Which equation of a line contains the point $(-5, 3)$ and is perpendicular to the graph of $2x - y = 3$?

   a. $y = -\frac{1}{2}x + \frac{1}{2}$
   b. $y = -2x + 5\frac{1}{2}$
   c. $y = -\frac{1}{2}x - \frac{1}{2}$
   d. $y = -2x - 5\frac{1}{2}$

7. Which statement best describes the relationship between the linear equations below?

$$2x - y = 8$$
$$4x - y = 8$$

   a. The linear equations are neither parallel nor perpendicular because they are the same line.
   b. The linear equations are parallel because their slopes are the same.
   c. The linear equations are perpendicular. Their slopes are opposite reciprocals of one another.
   d. The linear equations are not perpendicular. Their slopes are not opposite reciprocals of one another.

8. Which statement best describes the relationship between the linear equations below?

$$\frac{1}{2}x - y = 9$$
$$4x - 2y = 22$$

   a. The linear equations are perpendicular. Their slopes are opposite reciprocals of one another.
   b. The linear equations are not perpendicular. Their slopes are not opposite reciprocals of one another.
   c. The linear equations are neither parallel nor perpendicular because they are the same line.
   d. The linear equations are parallel because their slopes are equal to one another.

9. Which equation of a line contains the point $(4, -4)$ and is perpendicular to the graph of $4x + y = -2$?

   a. $y = \frac{1}{4}x - 3$
   b. $y = \frac{1}{4}x - 5$
   c. $y = 4x + 5$
   d. $y = 4x + 3$

10. Which statement best describes the relationship between the linear equations below?

$$\frac{3}{2}x + y = 4$$
$$-\frac{2}{3}x + y = 1$$

   a. The linear equations are parallel. The slope of both lines is $-\frac{2}{3}$.
   b. The linear equations are neither parallel nor perpendicular.
   c. The linear equations are perpendicular. The slope of both lines is $\frac{3}{2}$.
   d. The linear equations are perpendicular. Their slopes are opposite reciprocals of one another.

11. Which statement best describes the relationship between the linear equations below?

$$4x + y = 7$$
$$8x + 2y = 3$$

   a. The linear equations are parallel to one another. Their slopes are the same.
   b. The linear equations are neither parallel nor perpendicular to one another.
   c. The linear equations are not perpendicular. Their slopes are not opposite reciprocals of one another.
   d. The linear equations are perpendicular. Their slopes are opposite reciprocals of one another.

12. Which equation of a line contains the point $(0, -4)$ and is perpendicular to the graph of $3x + y = -12$?

   a. $y = -3x - 4$
   b. $y = 3x + 4$
   c. $y = \frac{1}{3}x - 4$
   d. $y = \frac{1}{3}x + 4$

13. When graphed on a coordinate plane, the red subway line has the equation $2x - 5y = 10$, and the blue subway line has the equation $5x + 2y = 12$. Determine whether these two subway lines are perpendicular to one another. Choose the statement that best describes their relationship.

   a. The subway lines are perpendicular. Their slopes are opposite reciprocals of one another.
   b. The subway lines are perpendicular. The slope of both lines is $\frac{2}{5}$.
   c. The subway lines are parallel. The slope of both lines is $-\frac{5}{2}$.
   d. The subway lines are neither parallel nor perpendicular.

14. When graphed on a coordinate plane, Broad Street has the equation $x + y = 3$. Market Street runs perpendicular to Broad Street and passes through the point $(-3, 2)$. What is the linear equation for Market Street?

   a. $y = x - 1$
   b. $y = x + 5$
   c. $y = -x + 1$
   d. $y = -x - 5$

15. When graphed on a coordinate plane, one side of a rectangle has the equation $2x - y = 2$. An adjacent side of the rectangle runs perpendicular to the first side and passes through the point $(-4, 0)$. What is the equation of the line for the adjacent side of the rectangle?

   a. $y = -\frac{1}{2}x + 2$
   b. $y = -\frac{1}{2}x - 2$
   c. $y = -2x + 2$
   d. $y = 2x - 2$

3.11 Calculating Correlation Coefficient

PRACTICE QUESTIONS

1. Given the data below, which statement best describes the correlation coefficient, rounded to the nearest hundredth, of the two variables?

x	64	73	70	75	63	66	69	70	75
y	140	190	185	190	165	160	160	170	190

   a. $r \approx -0.85$, there is a strong negative linear association between the two variables.
   b. $r \approx 0.85$, there is a strong positive linear association between the two variables.
   c. $r \approx 0.15$, there is a weak positive linear association between the two variables.
   d. $r = 0$, there is no association between the two variables.

2. Given the data below, which statement best describes the correlation coefficient, rounded to the nearest hundredth, of the two variables?

x	1	3	0.5	0.75	1.5
y	4	1	4	1.5	5

   a. $r = 0$, there is no correlation between the two variables.
   b. $r = -1$, there is a perfect negative linear association between the two variables.
   c. $r \approx -0.48$, there is a moderate negative linear correlation between the two variables.
   d. $r \approx 0.48$, there is a moderate positive linear correlation between the two variables.

3. Given the data below, which statement best describes the correlation coefficient, rounded to the nearest hundredth, of the two variables?

x	1	2	3	4	5	6	7	8	9
y	40	36	32	28	24	20	16	12	8

   a. $r \approx -0.92$, there is a strong negative linear association between the two variables.
   b. $r = 0$, there is no association between the two variables.
   c. $r = 1$, there is a perfect positive linear association between the two variables.
   d. $r = -1$, there is a perfect negative linear association between the two variables.

4. Given the data below, which statement best describes the correlation coefficient, rounded to the nearest hundredth, of the two variables?

x	25.5	22.5	30	26.5	34	29.5	25	26	26
y	25	30	36	20	23	28	25	16	14

   a. $r \approx 0.15$, there is a weak positive linear association between the two variables.
   b. $r \approx -0.15$, there is a weak negative linear association between the two variables.
   c. $r \approx 0.85$, there is a strong positive linear association between the two variables.
   d. $r = -1$, there is a perfect negative linear association between the two variables.

5. Given the data below, which statement best describes the correlation coefficient, rounded to the nearest hundredth, of the two variables?

x	1	1	2	2	3	3	4	4	5	5	6	6	7	7	8	8	9	9
y	3	7	2	6	2	3	8	4	7	3	4	8	2	3	2	6	3	7

   a. $r = 1$, there is a perfect positive linear association between the two variables.
   b. $r = 0$, there is no association between the two variables.
   c. $r = -1$, there is a perfect negative linear association between the two variables.
   d. $r \approx -0.09$, there is a weak negative linear association between the two variables.

6. Given the data below, which statement best describes the correlation coefficient, rounded to the nearest hundredth, of the two variables?

x	8	12	6	4	10	7	2	4	9
y	1	1	2	3	1	2	4	2	1

   a. $r \approx -0.50$, there is a moderate negative linear association between the two variables.
   b. $r \approx -0.89$, there is a strong negative linear association between the two variables.
   c. $r \approx 0.89$, there is a strong positive linear association between the two variables.
   d. $r = 0$, there is no association between the two variables.

7. Given the data below, which statement best describes the correlation coefficient, rounded to the nearest hundredth, of the two variables?

x	81	85	89	93	97	101	105	109	113
y	38.5	41	43.5	46	48.5	51	53.5	56	58.5

   a. $r \approx -0.46$, there is a moderate negative linear association between the two variables.
   b. $r = 0$, there is no linear association between the two variables.
   c. $r = 1$, there is a perfect positive linear association between the two variables.
   d. $r = -1$, there is a perfect negative linear association between the two variables.

8. Given the data below, which statement best describes the correlation coefficient, rounded to the nearest hundredth, of the two variables?

x	6	7	11	2	4	3	9	10	2
y	74	72	92	97	80	73	65	76	76

   a. $r \approx -0.14$, there is a weak negative linear association between the two variables.
   b. $r \approx 0.14$, there is a weak positive linear association between the two variables.
   c. $r = 0$, there is no linear association between the two variables.
   d. $r = 1$, there is a perfect positive linear association between the two variables.

9. Given the data below, which statement best describes the correlation coefficient, rounded to the nearest hundredth, of the two variables?

x	4	7	6	4	7	20	16	12
y	1	9	-1	6	10	19	20	-5

   a. $r = 1$, there is a perfect positive linear association between the two variables.
   b. $r = -1$, there is a perfect negative linear association between the two variables.
   c. $r \approx -0.63$, there is a moderate negative linear association between the two variables.
   d. $r \approx 0.63$, there is a moderate positive linear association between the two variables.

10. Given the data below, which statement best describes the correlation coefficient, rounded to the nearest hundredth, of the two variables?

x	5	5	7	6	5	7	4	3	8
y	8	4	13	11	5	13	3	2	12

a. $r \approx 0.92$, there is a strong positive linear association between the two variables.
b. $r \approx -0.81$, there is a strong negative linear association between the two variables.
c. $r \approx 0.18$, there is a weak positive linear association between the two variables.
d. $r = 0$, there is no association between the two variables.

11. Given the data below, which statement best describes the correlation coefficient, rounded to the nearest hundredth, of the two variables?

x	17	0	5	10	18	5	0	2	3
y	78	82	90	92	77	89	94	86	100

a. $r \approx -0.61$, there is a moderate negative linear correlation between the two variables.
b. $r \approx 0.61$, there is a moderate positive linear correlation between the two variables.
c. $r = 1$, there is a perfect positive linear association between the two variables.
d. $r = 0$, there is no correlation between the two variables.

12. Given the data below, which statement best describes the correlation coefficient, rounded to the nearest hundredth, of the two variables?

x	25	35	45	55	65	75	85	95	105
y	140	132	124	116	108	100	92	84	76

a. $r \approx -0.11$, there is a weak negative linear association between the two variables.
b. $r = 1$, there is a perfect positive linear association between the two variables.
c. $r = -1$, there is a perfect negative linear association between the two variables.
d. $r = 0$, there is no association between the two variables.

13. The table below shows how many hours each student spent studying for their final exam, as well as their final score on the exam out of 100%. Given this data set, which statement best describes the correlation coefficient, rounded to the nearest hundredth, of the two variables?

Name	Hours spent studying	Final exam score (out of 100%)
Abigail	1	60
Brian	1.5	80
Christina	2	75
David	3	80
Eloise	3.5	83
Flynn	4	90
Gianna	4.5	90
Hugo	5	85
Ira	5.5	90

  a. $r \approx -0.83$, there is a strong negative linear association between hours spent studying and the final exam score.
  b. $r \approx 0.84$, there is a strong positive linear association between hours spent studying and the final exam score.
  c. $r \approx 0.16$, there is a weak positive linear association between hours spent studying and the final exam score.
  d. $r = 0$, there is no association between hours spent studying and the final exam score.

14. The table below shows each student's height in inches, as well as the total number of books they read last year. Given this data set, which statement best describes the correlation coefficient, rounded to the nearest hundredth, of the two variables?

Name	Height (in inches)	Number of Books Read Last Year
Amma	68	20
Bella	66	22
Cathy	64	24
Dan	66	26
Eric	68	28

  a. $r \approx 0.54$, there is a moderate positive linear correlation between height and the number of books read last year.
  b. $r = 1$, there is a perfect positive linear association between height and the number of books read last year.
  c. $r = -1$, there is a perfect negative linear association between height and the number of books read last year.
  d. $r = 0$, there is no association between height and the number of books read last year.

15. The table below shows a paper company's total sales over the course of six years. Given this data set, which statement best describes the correlation coefficient, rounded to the nearest hundredth, of the two variables?

Year	Sales
2011	$80,000
2012	$76,000
2013	$77,000
2014	$74,000
2015	$70,000
2016	$72,000

a. $r = 1$, there is a perfect positive linear association between the year and total sales.
b. $r = -1$, there is a perfect negative linear association between the year and total sales.
c. $r \approx 0.91$, there is a strong positive linear association between the year and total sales.
d. $r \approx -0.91$, there is a strong negative linear association between the year and total sales.

3.12 Association, Correlation, and Causation

Practice Questions

1. Which of the following statements best describes a variable relationship that shows association but not causation?

a. The amount of flour needed for a cookie recipe and the total number of cookies the recipe yields
b. The amount of smoke and the size of a fire
c. The number of A's earned on a report card and overall grade point average
d. The number of cups of hot chocolate sold and pairs of mittens sold

2. Which of the following statements describes a variable relationship that best shows causation?

a. A secretary's typing speed in words per minute and the total amount of time it takes them to type 500 words
b. A toddler's age and the size of their vocabulary
c. The time it takes for a person to wash his or her hair and the total amount of shampoo needed
d. A middle school student's height and their score on a proficiency test

3. Which of the following statements best describes a variable relationship that shows association but not causation?

a. The amount of time spent exercising on a rowing machine and the number of calories burned while working out
b. Ounces of soup and the total number of servings
c. A sprinter's running speed in minutes per mile and the total amount of time it takes for them to run 1 mile
d. The day of the week and the number of airline tickets sold

4. Which of the following statements describes a variable relationship that best shows causation?

a. The unit rate of 1 pencil and the total cost of a box of 100 pencils
b. A person's age and the number of countries they've traveled to
c. A person's shoe size and their shirt size
d. The number of yards a football team gains on offense and the number of touchdowns scored

5. Which of the following statements describes a variable relationship that shows association but not causation?

a. The number of hours spent babysitting and the total amount of money paid for the babysitting job
b. A student's high school attendance record and their SAT score
c. The number of books purchased and the total amount of money spent at the bookstore
d. The number of steps taken and the total number of miles walked in a day

6. Which of the following statements describes a variable relationship that shows association but not causation?
   a. A person's speed on a stationary bike and the total amount of time it takes for them to bike 10 miles
   b. The number of hours of television a student watches per week and the number of homework assignments completed
   c. The height of a child and their clothing size
   d. The cost of bulk candy by the pound and the total amount spent on 5 pounds of candy

7. Which of the following statements describes a variable relationship that best shows causation?
   a. The number of years of education a person completes and their earning potential
   b. The number of sodas a person consumes each day and their body mass index
   c. The number of pages in a photo album and the total number of pictures it can hold
   d. The number of hours a student spends studying for a test and their test score

8. Which of the following statements describes a variable relationship that best shows causation?
   a. The dimensions of a restaurant's outdoor patio and the total amount of square footage available on the patio for dining
   b. The number of days a heating system runs throughout the year and the outdoor temperature
   c. A person's age and their total amount of retirement savings
   d. The total number of books a person has read and their score on an IQ test

9. Which of the following statements describes a variable relationship that shows association but not causation?
   a. The percentage of battery power on a laptop and the total amount of time before it shuts down
   b. The square footage of a person's home and their annual income
   c. A person's walking speed and the total amount of time it takes for them to walk 3 miles
   d. The cost of gas per gallon and the total amount spent to fill up a 12-gallon gas tank

10. Which of the following statements describes a variable relationship that best shows causation?
    a. A student's SAT score and their college grade point average
    b. The amount of time a person spends exercising and their total weight loss
    c. The amount of gasoline put into a car and the distance it can travel
    d. A company's annual sales and its total revenue

11. Which of the following statements describes a variable relationship that best shows causation?
    a. An athlete's weight and their running speed
    b. The amount of force exerted on a ball and the distance it travels over a flat surface
    c. The number of cups of lemonade sold and the number of popsicles sold
    d. A child's age and the number of states they have traveled to

12. Which of the following statements describes a variable relationship that shows association but not causation?
   a. The number of extra treats a dog consumes and its weight gain
   b. A student's number of correctly spelled words on a spelling test and their overall score
   c. The number of cups of coffee consumed and the amount of time spent awake
   d. The amount of wind present and the size of the waves in a body of water

13. Dave notices that the older he gets, the more time he has available to read. He wants to know if there is a relationship between a person's age and the total number of books they've read. Which of the following statements most likely best explains the relationship between these two variables?
   a. There is a positive correlation and causation in the relationship between a person's age and the total number of books they've read.
   b. There is no causation or association between a person's age and the total number of books they've read.
   c. There is a positive correlation between a person's age and the total number of books they've read but this is not a causal relationship.
   d. There is causation but no association between a person's age and the total number of books they've read.

14. Brooke is buying a fish tank for her aquarium. She wants to know if there is a relationship between the size of a fish tank and the number of gallons of water it can hold. Which of the following statements best explains the relationship between these two variables?
   a. There is a positive correlation and causation in the relationship between the size of a fish tank and the number of gallons of water it can hold.
   b. There is a negative correlation between the size of a fish tank and the number of gallons of water it can hold but this is not a causal relationship.
   c. There is a positive correlation between the size of a fish tank and the number of gallons of water it can hold but this is not a causal relationship.
   d. There is no causation or association between the size of a fish tank and the number of gallons of water it can hold.

15. Ben is working on his free throw percentage. He wants to know if there is a relationship between the number of free throw shots taken and the total number of points scored. Which of the following statements best explains the relationship between these two variables?
   a. There is no causation or association between the number of free throw shots taken and the total number of points scored.
   b. There is a positive correlation and causation in the relationship between the number of free throw shots taken and the total number of points scored.
   c. There is causation but no association between the number of free throw shots taken and the total number of points scored.
   d. There is a positive correlation between the number of free throw shots taken and the total number of points scored but this is not a causal relationship.

Chapter 4: Inequalities

4.1 Inequality Basics

Practice Questions

1. Which inequality symbol means less than?

    a. $>$
    b. $\leq$
    c. $<$
    d. $\geq$

2. Translate the phrase below into an algebraic inequality:

    x is less than or equal to 4

    a. $x \geq 4$
    b. $x < 4$
    c. $x > 4$
    d. $x \leq 4$

3. Translate the algebraic inequality below into a phrase:

    $x \geq 10$

    a. x is greater than 10
    b. x is less than 10
    c. x is greater than or equal to 10
    d. x is less than or equal to 10

4. What is the meaning of the inequality symbol $>$?

    a. Less than
    b. Greater than
    c. Less than or equal to
    d. Greater than or equal to

5. Which number line shows the solution set for $x > 8$?

    a.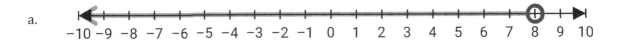

    b.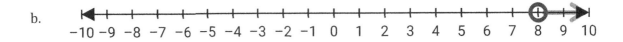

    c.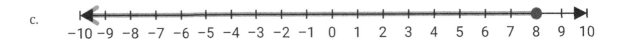

d.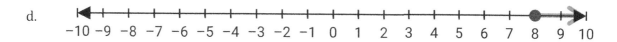

6. Which inequality statement matches the solution set graphed on the number line below?

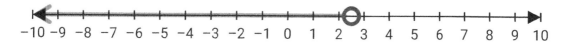

a. $x < 2.5$
b. $x > 2.5$
c. $x \leq 2.5$
d. $x \geq 2.5$

7. Which inequality symbol means greater than or equal to?

a. $>$
b. $<$
c. $\geq$
d. $\leq$

8. Translate the phrase below into an algebraic inequality:

x is less than 3

a. $x \geq 3$
b. $x \leq 3$
c. $x > 3$
d. $x < 3$

9. Which number line shows the solution set for $x \leq$?

a.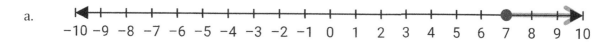

b.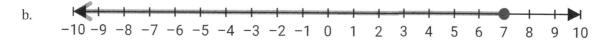

c.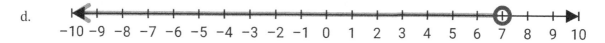

d.

10. What is the meaning of the inequality symbol $\leq$?

a. Greater than
b. Less than
c. Greater than or equal to
d. Less than or equal to

11. Translate the algebraic inequality below into a phrase:

$$x \geq -2$$

   a. x is greater than or equal to -2
   b. x is greater than -2
   c. x is less than or equal to -2
   d. x is less than -2

12. Which inequality statement matches the solution set graphed on the number line below?

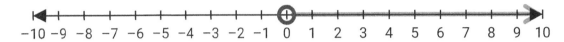

   a. $x < 0$
   b. $x > 0$
   c. $x \leq 0$
   d. $x \geq 0$

13. Jane is analyzing her finances. She wants to have at least $200 in her savings account, and she makes a note of this by writing an inequality where x represents the amount of money in her savings account. Which inequality best represents this statement?

   a. $x \leq 200$
   b. $x \geq 200$
   c. $x < 200$
   d. $x > 200$

14. Breanna wants to go to an arcade with her friends. Breanna's mom tells her that she can go if she agrees to spend no more than 20 dollars. Which number line best represents the solution set for this scenario?

   a.

   b.

   c.

   d.

15. Cathy is looking at the weather forecast for the upcoming week. Saturday's average temperature will be colder than 65 °F. Which inequality best represents Saturday's temperature, x?

   a. $x \leq 65$
   b. $x \geq 65$
   c. $x < 65$
   d. $x > 65$

4.2 Solving One-Step Inequalities

PRACTICE QUESTIONS

1. Solve the inequality:
$$x + 8 < 12$$

   a. $x < 4$
   b. $x < 20$
   c. $x > 4$
   d. $x > -4$

2. Solve the inequality:
$$x - 15 > 3$$

   a. $x > 18$
   b. $x > 12$
   c. $x = 18$
   d. $x > -18$

3. Solve the inequality:
$$5x \leq 70$$

   a. $x \leq 65$
   b. $x \leq 350$
   c. $x \geq -14$
   d. $x \leq 14$

4. Solve the inequality:
$$\frac{x}{3} \geq 6$$

   a. $x \leq -18$
   b. $x \geq 9$
   c. $x \geq 2$
   d. $x \geq 18$

5. Solve the inequality:
$$x - 2 < 12.5$$

   a. $x < 10.5$
   b. $x < 14.5$
   c. $x \leq 14.5$
   d. $x < 25$

6. Solve and graph the inequality:
$$7x > 133$$

a. $x < -19$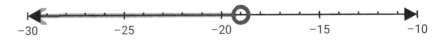

b. $x > 140$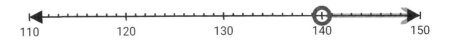

c. $x > 19$

d. $x > 125$

7. Solve the inequality:
$$\frac{x}{12} \leq 2$$

a. $x \leq 6$
b. $x \leq 24$
c. $x \leq 14$
d. $x \leq 10$

8. Solve and graph the inequality:
$$x + 9.25 \geq 21.25$$

a. $x \leq 30.5$

b. $x \leq 12$

c. $x \geq 30.5$

d. $x \geq 12$

9. Solve the inequality:
$$-4x < 52$$

a. $x \geq -208$
b. $x < -208$
c. $x > -13$
d. $x < -13$

10. Solve the inequality:
$$\frac{x}{18} > 11$$

a. $x > 29$
b. $x > -7$
c. $x > 198$
d. $x > \frac{11}{18}$

11. Solve and graph the inequality:
$$x + 54 \leq -3$$

a. $x \geq 57$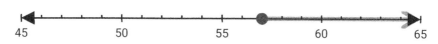

b. $x \leq -57$

c. $x \leq 51$

d. $x \geq -51$

12. Solve the inequality:
$$x - 15 \geq 22$$

a. $x < 7$
b. $x < 37$
c. $x \geq 7$
d. $x \geq 37$

13. Stephanie baked a batch of cookies for her friends. She split the cookies into three equal groups and noted that each friend will receive more than a dozen cookies. Based on this information, choose the inequality statement and graph that represent the total number of cookies Stephanie could have baked.

a. $x \geq 36$

b. $x > 36$

c. $x < 36$

d. $x \leq 36$

14. Alex's lemonade stand must sell more than 45 cups of lemonade to make a profit. He has already sold 30 cups. Based on this information, which inequality best represents the number of cups of lemonade, x, Alex still needs to sell?

a. $x > 15$
b. $x < 15$
c. $x \geq 15$
d. $x \leq 15$

15. Amaya went to the gas station and filled up her car's 12-gallon gas tank. Her gas tank uses 4 gallons of gas per hour when driving on the interstate. Based on this information, which inequality statement and graph best represents the number of hours, x, Amaya can drive on the interstate before she runs out of gas.

a. $x \geq 3$

b. $0 \leq x \leq 3$

c. $x > 3$

d. $x < 3$

4.3 Solving Multi-Step Inequalities

PRACTICE QUESTIONS

1. Solve the inequality: $9x + 15 \leq -12$.
   a. $x \leq -3$
   b. $x \geq 3$
   c. $x \leq \frac{1}{3}$
   d. $x \leq -\frac{1}{3}$

2. Solve the inequality: $5x + 6 > 14 + 3x$.
   a. $x < -4$
   b. $x > 4$
   c. $x < -8$
   d. $x > 3$

3. Solve the inequality: $3 - 2x < 13 - 7x$.
   a. $x < 2$
   b. $x > -2$
   c. $x > -10$
   d. $x < 1$

4. Solve the inequality: $2(x - 5) \geq 6x + 18$.
   a. $x \geq -\frac{28}{5}$
   b. $x \leq -\frac{13}{4}$
   c. $x \geq 7$
   d. $x \leq -7$

5. Solve the inequality: $x - 18 < -4(x - 3)$.
   a. $x > 15$
   b. $x < 3$
   c. $x < 6$
   d. $x > -6$

6. Solve the inequality: $15 < -6(x + 1) - x$.
   a. $x > \frac{21}{2}$
   b. $x < -\frac{7}{3}$
   c. $x < -3$
   d. $x > 3$

7. Solve the inequality: $6x + 4(2x - 1) > 2 + 3(x + 9)$.
   a. $x < -3$
   b. $x > 1$
   c. $x < -4$
   d. $x > 3$

8. Graph the solution to the inequality: $7x - 8 > 2x + 17$.

a.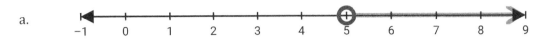

b.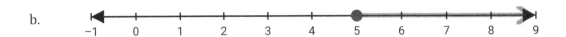

c.

d.

9. Graph the solution to the inequality: $4 - 3(x + 5) \leq 4x + 3$.

a.

b.

c.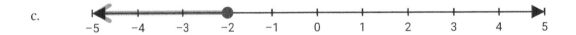

d.

10. Graph the solution to the inequality: $14 - 2x < -2(5x + 9)$.

a.

b.

c.

d.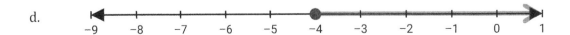

11. Graph the solution to the inequality: $2(x-3) \geq 4x+8$.

a.

b.

c.

d.

12. Solve and graph the following inequality: $2x - 1 < 7$.

a.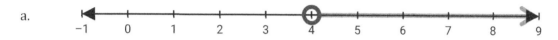

b.

c.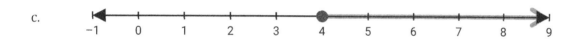

d.

13. The cost, in dollars, for a company to make x number of calculators can be represented by the expression $25x + 1,000$. If the cost is not to exceed $2,500, what is the maximum number of calculators that can be made?
   a. 90
   b. 100
   c. 140
   d. 60

14. Twelve more than three times a real number, x, is at least as large as the sum of four times the number and ten. Which of the following statements must be true about the real number?
    a. The real number must be less than or equal to 2.
    b. The real number must be greater than or equal to 2.
    c. The real number must be strictly less than 2.
    d. The real number must be strictly greater than 2.

15. The perimeter of a rectangle is less than or equal to 80 meters. If the length of the rectangle is four more than twice its width, what is the maximum the width can be?
    a. 10 meters
    b. 6 meters
    c. 12 meters
    d. 14 meters

4.4 Solving Compound Inequalities

PRACTICE QUESTIONS

1. Solve the compound inequality: $3x + 5 > 11$ or $2x + 3 \leq -1$.
    a. $x \leq -2$ or $x \leq 2$
    b. $x > -2$ or $x \leq 2$
    c. $x \leq -2$ or $x > 2$
    d. $x \geq -2$ or $x < 2$

2. Solve the compound inequality: $4x - 6 > 18$ and $3x - 10 \leq 20$.
    a. $6 < x \leq 10$
    b. $6 \leq x < 10$
    c. $10 \leq x < 6$
    d. $6 > x \geq 10$

3. Solve the combined inequality: $5x - 8 \leq 17$ or $2x + 4 \geq -2$.
    a. $x \leq -3$ or $x \leq 5$
    b. $x \geq 5$ or $x \leq -3$
    c. $x \leq -3$ or $x \geq 5$
    d. $x \leq 5$ or $x \geq -3$

4. Solve the compound inequality: $21 < 6x - 9$ and $-4x + 1 > -35$.
    a. $5 < x < 9$
    b. $9 < x < 5$
    c. $5 > x > 9$
    d. $9 > x < 5$

5. Solve the compound inequality: $-3x + 5 > 12$ or $5 \leq 2x + 6$.
    a. $x < -\frac{7}{3}$ or $x \leq -\frac{1}{2}$
    b. $x < -\frac{7}{3}$ or $x \geq -\frac{1}{2}$
    c. $x > -\frac{7}{3}$ or $x \leq -\frac{1}{2}$
    d. $x > -\frac{7}{3}$ or $x \geq -\frac{1}{2}$

6. Solve the compound inequality: $6x + 9 \geq 20$ and $2x - 3 \leq -14$.
    a. $\frac{11}{6} \leq x \leq -\frac{11}{2}$
    b. $-\frac{11}{2} \leq x \leq \frac{11}{6}$
    c. All real numbers
    d. No solution

7. Solve the compound inequality: $-4x + 11 \geq 27$ or $10 > -x + 17$.
    a. $x \leq -4$ or $x < 7$
    b. $x \geq 4$ or $x < -7$
    c. $x \leq -4$ or $x > 7$
    d. $x \geq -4$ or $x < 7$

8. Solve the compound inequality: $13 < -3x + 19$ and $4 \geq 1 - 3x$.
   a. $-1 < x \leq 2$
   b. $-1 \leq x < 2$
   c. $-1 \geq x > 2$
   d. $-1 > x < 2$

9. Graph the solution to the compound inequality: $2x + 1 < 5$ or $x + 2 > 7$.

   a.

   b.

   c.

   d.

10. Graph the solution to the compound inequality: $4x + 3 \geq -9$ and $2x - 9 < 3$.

    a.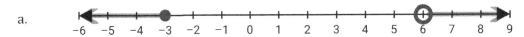

    b.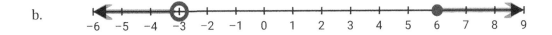

    c.

    d.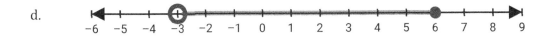

11. Graph the solution to the compound inequality: $-2x - 5 > 13$ or $9 - 6x \leq -3$.

a. [number line from -15 to 10, open circle at -9, shaded left; closed circle at 2, shaded right]

b. [number line from -15 to 10, closed circle at -9, shaded left; open circle at 2, shaded right]

c. [number line from -15 to 10, closed circle at -9, shaded right to open circle at 2]

d. [number line from -15 to 10, open circle at -9, shaded right to closed circle at 2]

12. Graph the solution to the compound inequality: $3x - 5 \geq 7$ and $-9 < 2x + 1$.

a. [number line from -7 to 8, closed circle at -5 shaded right to open circle at 4]

b. [number line from -7 to 8, open circle at -5, shaded right to closed circle at 4]

c. [number line from -7 to 8, open circle at -5, shaded right]

d. [number line from -7 to 8, closed circle at 4, shaded right]

13. A gardener is building a rectangular garden. The perimeter of the garden can be no more than 70 feet and no less than 60 feet. If the length of the garden must be 25 feet, what is the range of values the width can be in feet?

a. $5 \geq x \geq 10$
b. $5 \leq x \leq 10$
c. $10 > x < 5$
d. $5 \geq x \leq 10$

14. The sum of three times a real number, x, and fifteen is either less than or equal to forty-five or greater than ninety. What is the range of values the number can be?

a. $x \leq 10$ or $x < 25$
b. $x \geq 25$ or $x < 10$
c. $x \leq 10$ or $x > 25$
d. $x \geq 10$ or $x < 25$

15. The total cost, C, that a company incurs for producing x number of ear buds in dollars is:
$$C(x) = 10x + 50$$
If the company can afford a cost that is no less than $1,000 and no more than $7,000, what is the range of the number of ear buds it can afford to produce?

   a. $95 \leq x \leq 695$
   b. $95 < x < 695$
   c. $95 > x \leq 695$
   d. $95 \geq x \geq 695$

4.5 Solving Inequalities Involving Absolute Values

PRACTICE QUESTIONS

1. Solve the inequality $|x| < 4$ and graph the solution on a number line, if possible.

a. $-4 < x < 4$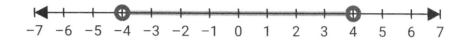

b. $x < -4$ or $x > 4$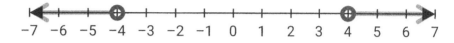

c. All real numbers

d. This inequality has no solution.

2. Solve the inequality $|x| \geq 3$ and graph the solution on a number line, if possible.

a. $-3 \leq x \leq 3$

b. $x \leq -3$ or $x \geq 3$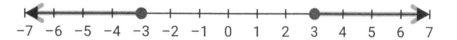

c. All real numbers

d. This inequality has no solution.

3. Solve the inequality $|x| < -2$ and graph the solution on a number line, if possible.

a. $-2 < x < 2$

b. $x < -2$ or $x > 2$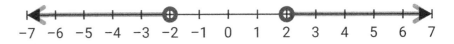

c. All real numbers

d. This inequality has no solution.

4. Solve the inequality $|x| + 5 > 6$.
   a. $x < -11$ or $x > 11$
   b. $x < -1$ or $x > 1$
   c. $-11 < x < 11$
   d. $-1 < x < 1$

5. Solve the inequality $|x| - 3 \leq -1$ and graph the solution on a number line, if possible.

a. $-2 \leq x \leq 2$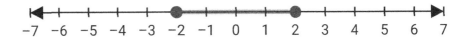

b. $x \leq -2$ or $x \geq 2$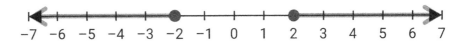

c. $-4 \leq x \leq 4$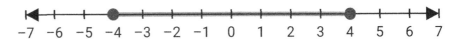

d. This inequality has no solution.

6. Solve the inequality $-4|x| \leq -6$.
   a. $-2 \leq x \leq 2$
   b. $-\frac{3}{2} \leq x \leq \frac{3}{2}$
   c. $x \leq -\frac{3}{2}$ or $x \geq \frac{3}{2}$
   d. This inequality has no solution.

7. Solve the inequality $2|x| - 7 \leq -1$.
   a. $-\frac{13}{2} \leq x \leq \frac{13}{2}$
   b. $x \leq -\frac{13}{2}$ or $x \geq \frac{13}{2}$
   c. $-3 \leq x \leq 3$
   d. $x \leq -3$ or $x \geq 3$

8. Solve the inequality $-3|x| + 1 \leq 13$.
   a. $-4 \leq x \leq 4$
   b. $x \leq -4$ or $x \geq 4$
   c. $x \leq -\frac{15}{3}$ or $x \geq \frac{15}{3}$
   d. All real numbers

9. Solve the inequality $|x - 3| > 1$ and graph the solution on a number line if possible.

   a. $2 < x < 4$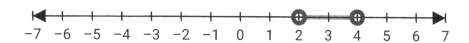

   b. $x < 2$ or $x > 4$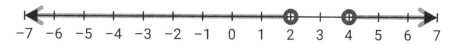

   c. $x > 4$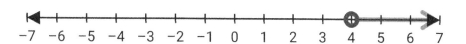

   d.   All real numbers

10. Solve the inequality $|4x + 12| < 8$.
    a. $x < -14$ or $x > -10$
    b. $-14 < x < -10$
    c. $x < -5$ or $x > -1$
    d. $-5 < x < -1$

11. Solve the inequality $|x - 1| \leq 3$.
    a. $x \leq -2$ or $x \geq 4$
    b. $-2 \leq x \leq 4$
    c. $x \leq 2$
    d. $-4 \leq x \leq 2$

12. Solve the inequality $|2x + 1| > 6$.
    a. $-\frac{7}{2} < x < \frac{5}{2}$
    b. $x < -\frac{7}{2}$ or $x > \frac{5}{2}$
    c. $-4 < x < 2$
    d. $x < -4$ or $x > 2$

13. The food safety guidelines at a restaurant specify that if food is being held for serving, the temperature of the food must differ from 90 °F by more than 50 °F, whether higher or lower. Using x as the holding temperature, set up and solve an inequality with an absolute value to determine what holding temperatures are allowed.
    a. $x < -40$ °F or $x > 140$ °F
    b. -40 °F $< x < 140$ °F
    c. $x < 40$ °F or $x > 140$ °F
    d. 40 °F $< x < 140$ °F

14. A statistician says he has collected data showing that the mean height of men (adult males) in the United States in the late twentieth century was 69 inches and that ninety-five percent of men had heights within 5.8 inches of this mean. Using x as a man's height, set up and solve an inequality with an absolute value to determine the heights of ninety-five percent of men in the United States at this time.
    a. $x \leq 65.55$ inches or $x \geq 72.45$ inches
    b. $x \leq 63.2$ inches or $x \geq 74.8$ inches
    c. 65.55 inches $\leq x \leq 72.45$ inches
    d. 63.2 inches $\leq x \leq 74.8$ inches

15. A theme park has a water slide in which groups of riders ride small boats from the top of the slide to the bottom. The slide operator has a large scale at the top of the slide to weigh each group of riders before they enter the boat. If a group is too heavy or too light, then the boat is unstable. So, the weight of each group must not differ from the ideal weight of 650 lb. by more than 120 lb. Using x as the weight of a group, set up and solve an inequality with an absolute value to determine the weights that are safe.
    a. -120 lb $\leq x \leq 120$ lb
    b. 0 lb $\leq x \leq 120$ lb
    c. $x \leq 530$ lb or $x \geq 770$ lb
    d. 530 lb $\leq x \leq 770$ lb

4.6 Writing Linear Inequalities

Practice Questions

1. Which inequality statement can be represented by the graph shown below?

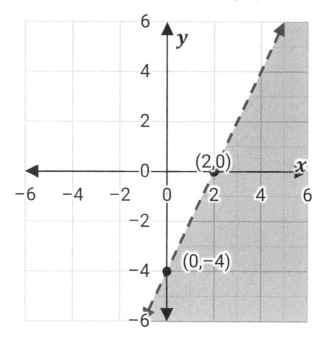

   a. $y < 2x - 4$
   b. $y < 2x + 4$
   c. $y < -2x - 4$
   d. $y \leq 2x - 4$

2. The drama club at a school is putting on a play to raise money. They hope to raise $520 to cover the cost of the play and the end of school trip. An adult ticket costs $6, and a child ticket costs $2. Which inequality statement represents all the possible combinations of adult tickets, x, and child tickets, y, that will help them raise enough money?

   a. $6x + 2y \geq 520$
   b. $6x + 2y \leq 250$
   c. $2x + 6y \geq 520$
   d. $2x + 6y \leq 520$

3. A company creates paint by number kits. The company starts with a $400 investment and earns $25 from each sale they make. If they need to operate on less money than they have, which of the following inequalities represents the amount of money they have to operate from, y, based on the number of kits they sell, x?

   a. $25x + y > 400$
   b. $y > -25x + 400$
   c. $25x - y < 400$
   d. $y < 25x + 400$

4. Which inequality statement can be represented by the graph shown below?

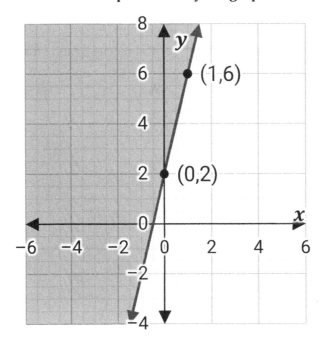

a. $y \geq 4x + 2$
b. $y > 4x + 2$
c. $y \geq -4x + 2$
d. $y \geq 4x - 2$

5. Sarah has a food delivery service. She charges $10 for delivery plus 50 cents per mile and allows for customers to give her tips. Which of the following inequalities represents the amount of money she earns, y, for a delivery that is x miles away?

a. $y < 0.5x + 10$
b. $y > 10x - 0.5$
c. $y \geq 0.5x + 10$
d. $y \leq 10x + 0.5$

6. Which inequality statement can be represented by the graph shown below?

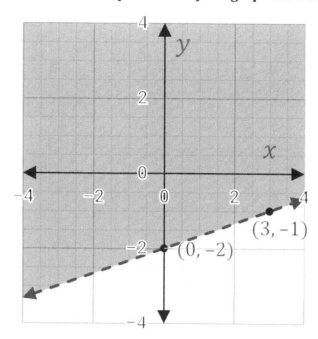

a. $y \leq \frac{1}{3}x - 2$
b. $y > \frac{1}{3}x - 2$
c. $y > 3x - 2$
d. $y > \frac{1}{3}x + 2$

7. Jackson runs a lawn mowing business where he earns $19 per hour he spends mowing lawns. Some weeks there are too many lawns for him to mow on his own, so he pays his brother $14 per hour to help him. If Jackson has never earned more than $600 in a week, which of the following inequalities represents the combination of hours that he has mowed, x, and the hours his brother has mowed, y?

a. $19x + 14y \geq 600$
b. $19x - 14y \leq 600$
c. $-19x + 14y < 600$
d. $19x - 14y > 600$

8. Which inequality statement can be represented by the graph shown below?

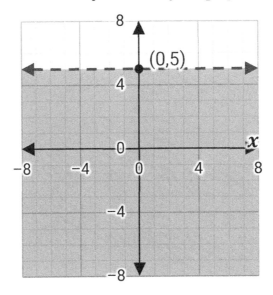

a. $y < 5$
b. $y > 5$
c. $y < -5$
d. $x < 5$

9. Which inequality statement can be represented by the graph shown below?

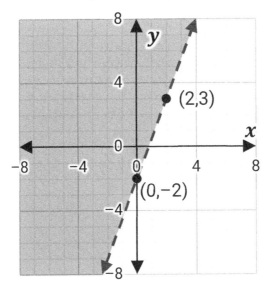

a. $y \leq \frac{5}{2}x + 2$
b. $y > \frac{5}{2}x + 2$
c. $y \geq \frac{5}{2}x - 2$
d. $y > \frac{5}{2}x - 2$

10. Which inequality statement can be represented by the graph shown below?

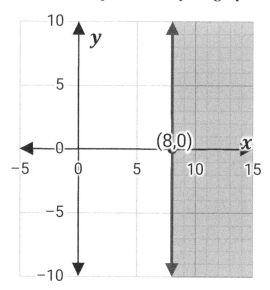

a. $x < 8$
b. $x > 8$
c. $x \geq 8$
d. $x \leq 8$

11. Which inequality statement can be represented by the graph shown below?

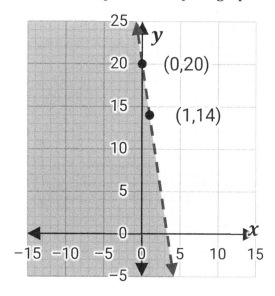

a. $y < 6x - 20$
b. $y < -6x + 20$
c. $y \leq -6x - 20$
d. $y \leq 6x + 20$

12. Which inequality statement can be represented by the graph shown below?

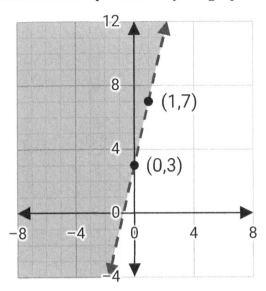

   a. $y < 3x - 4$
   b. $y > 3x + 4$
   c. $y \geq 4x + 3$
   d. $y > 4x + 3$

13. Kelly is at a basketball game with her friends. She plans to buy some hotdogs and sodas for the group. Each hotdog costs 2 dollars, and each soda costs 4 dollars. Kelly wants to know how many hotdogs and sodas she can buy with 40 dollars. Which inequality statement represents all possible combinations of the number of hotdogs, x, and the number of sodas, y, that Kelly can buy for no more than 40 dollars?

   a. $4x + 40y \geq 2$
   b. $40x - 2y > 4$
   c. $2x + 4y \leq 40$
   d. $2x - 4y < 40$

14. Glenn is visiting Philadelphia and wants to rent a bicycle. The bike costs $12 for a daily rental plus $0.15 for each minute of use. Glenn wants to rent a bicycle for as long as he can afford, but he can't spend more than y dollars. Which inequality statement represents the maximum number of minutes, x, Glenn can rent the bicycle for in terms of y dollars?

   a. $y \leq 0.15x + 12$
   b. $y > 0.15x + 12$
   c. $y < 12x - 0.15$
   d. $y \geq 12x - 0.15$

15. A coffee shop is getting new tables. Each small table seats three people, and each large table seats six people. The owner of the coffee shop wants to know how many small and large tables she should purchase. The maximum capacity of the coffee shop is 30 people. Which inequality statement represents all possible combinations of the number of large tables, x, and the number of small tables, y, that the coffee shop owner can seat for a maximum of 30 people?

 a. $3x - 30y \geq 6$
 b. $30x + 6y < 3$
 c. $6x + 3y \leq 30$
 d. $6x + 3y > 30$

4.7 Graphing Linear Inequalities

PRACTICE QUESTIONS

1. Graph the inequality: $3x + 2y \geq 12$.

a.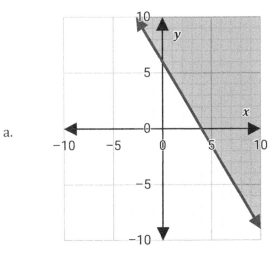

b.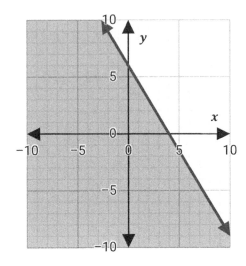

c.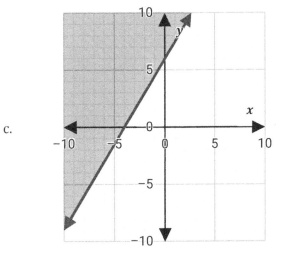

d.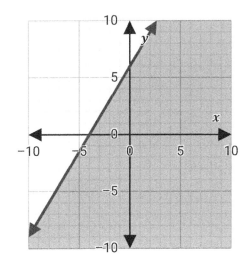

2. Graph the inequality: $x + 3y \leq 6$.

a.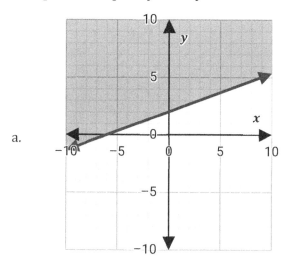

b.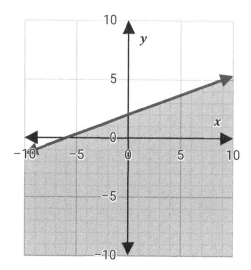

c.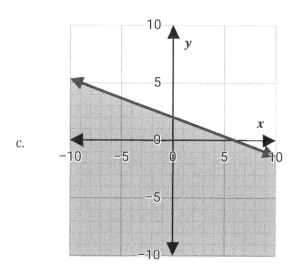

d.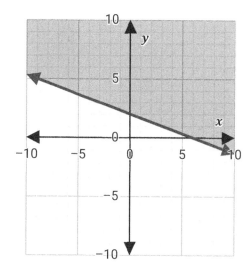

3. Solve the inequality: $5x - 10y < 4$.

a.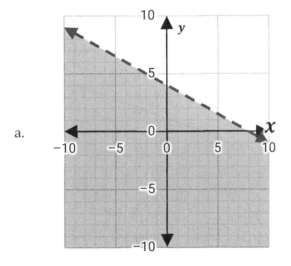

b.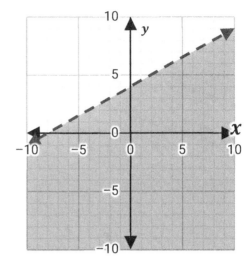

c.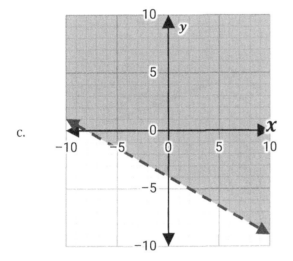

d.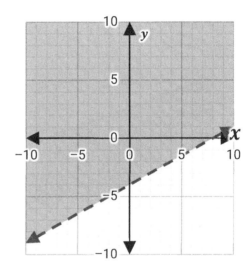

4. Graph the inequality: $\frac{1}{3}x + \frac{1}{2}y >$.

a.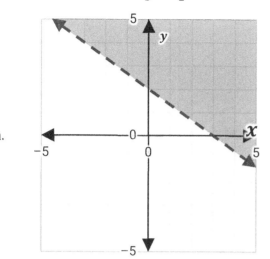

b.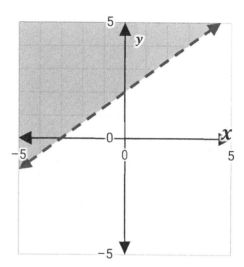

c.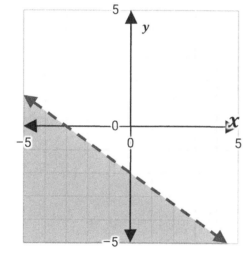

d.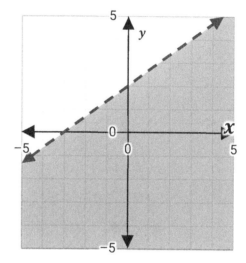

5. Graph the inequality: $2x - \frac{2}{3}y \leq 1$.

a.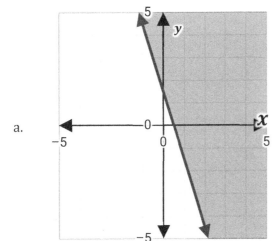

b.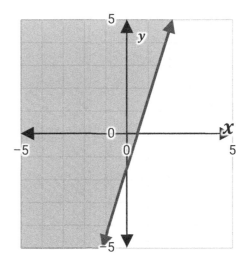

c.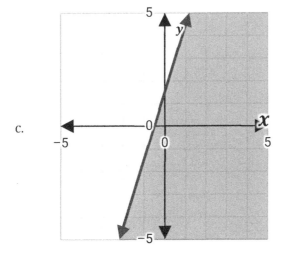

d.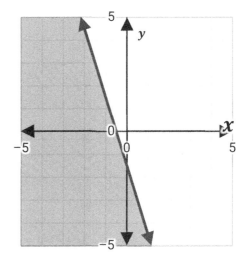

6. Graph the inequality: $-3x + \frac{3}{4}y < 6$.

a.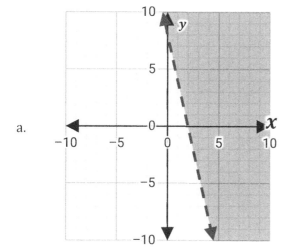

b.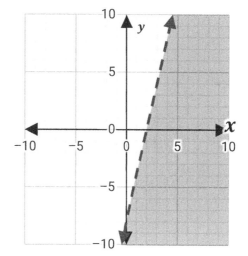

c.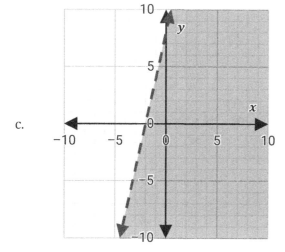

d.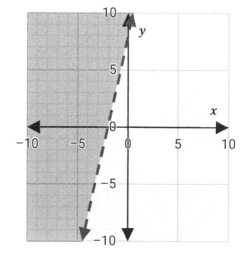

7. **Graph the inequality:** $y > \frac{2}{3}x + 1$.

a.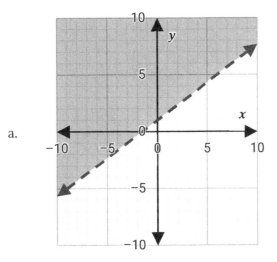

b.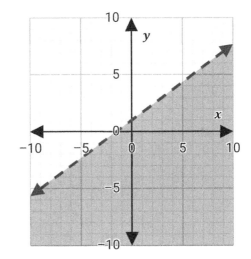

c.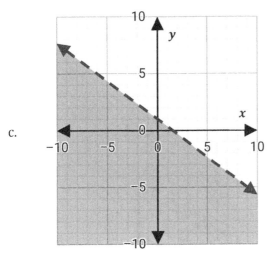

d.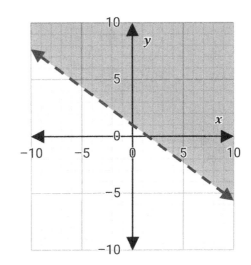

8. Graph the inequality: $y \leq \frac{3}{5}x - 2$.

a.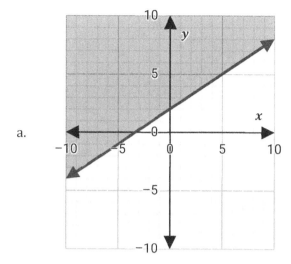

b.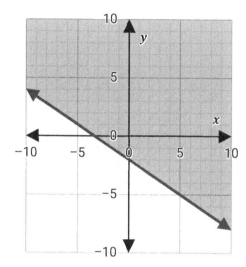

c.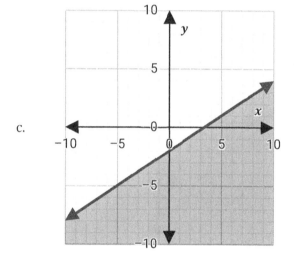

d.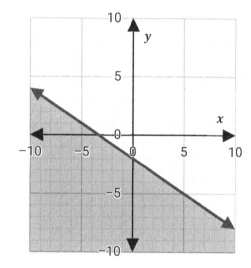

9. Graph the inequality: $y \geq -\frac{4}{7}x + 2$.

a.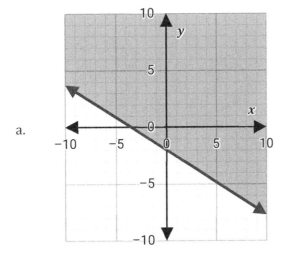

b.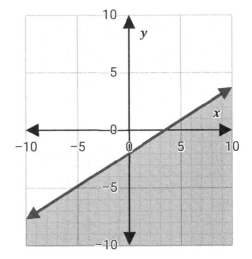

c.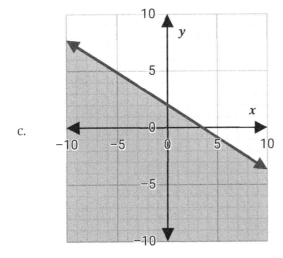

d.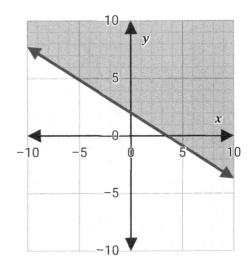

10. Graph the inequality: $y \leq -\frac{5}{2}x - 3$.

a.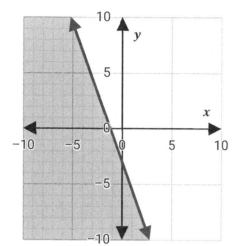

b.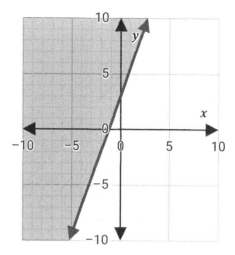

c.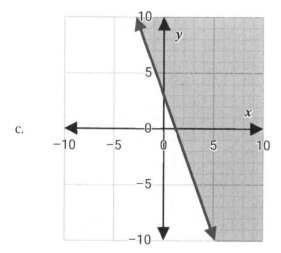

d.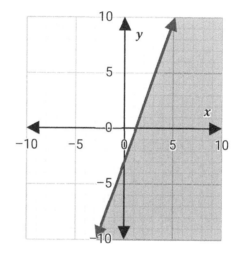

11. Graph the inequality: $y \leq 4x$.

a.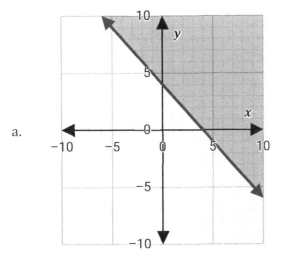

b.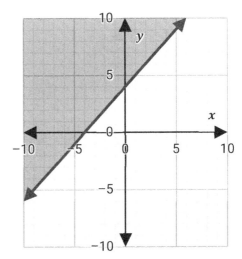

c.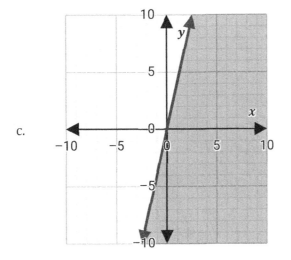

d.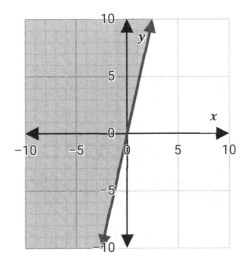

12. Graph the inequality: $y < -3x + 4$.

a.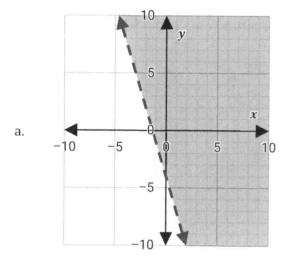

b.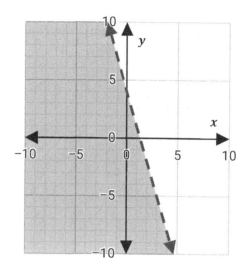

c.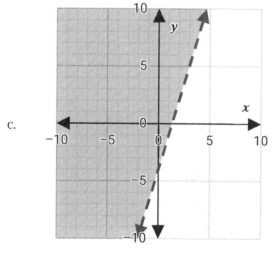

d.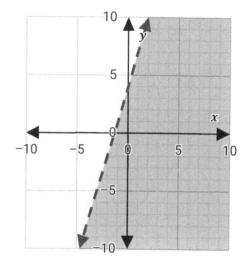

13. There are x number of male employees and y number of female employees that work for a company. If the company can have no more than 65 employees working for it at any given time, which graph represents the combination for the number male and female employees that work for the company?

a.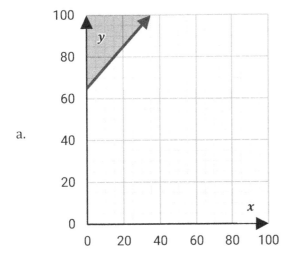

b.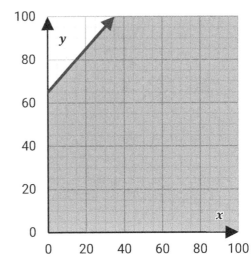

c.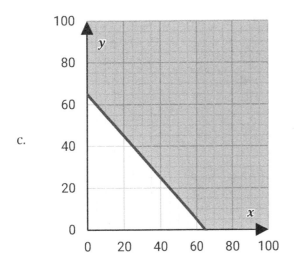

d.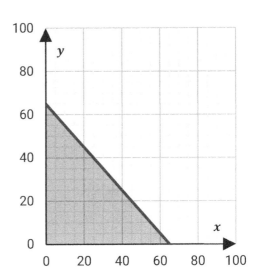

14. A department store purchases shirts for $5.00 each and purchases a pair of pants for $10.00 each. If x represents the number of shirts the store purchases and y represents the number of pants the store purchases, which graph represents the combination of shirts and pants the store can purchase so the total cost does not exceed $500.00?

a.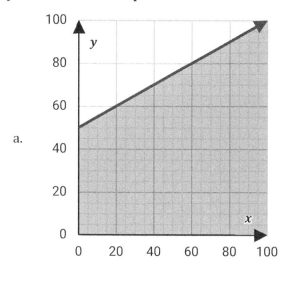

b.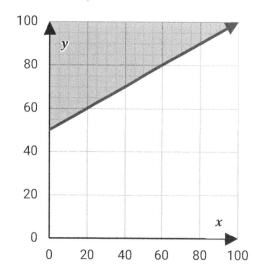

c.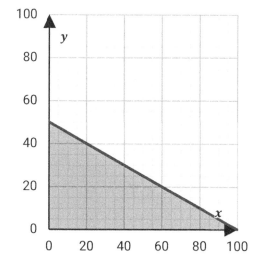

d.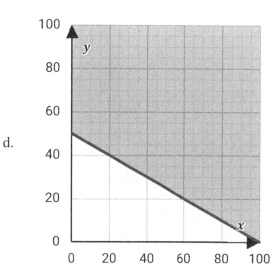

15. Let x and y be two numbers. If y is greater than five more than three times x, which graph represents the different combinations for x and y?

a.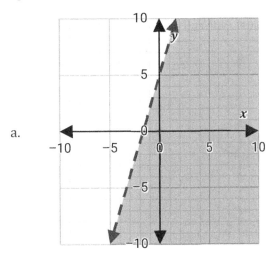

b.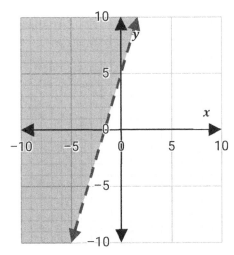

c.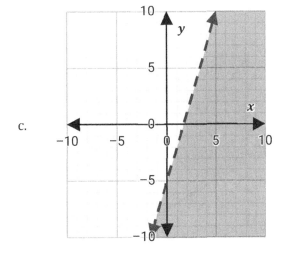

d.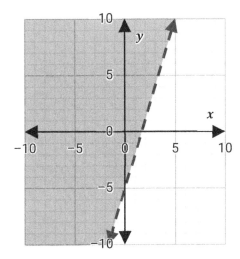

Chapter 5: Systems of Equations

5.1 The Graphing Method

PRACTICE QUESTIONS

1. Solve the following system of equations using the graphing method: $\begin{cases} y = -\frac{1}{2}x + 4 \\ y = x + 1 \end{cases}$.

a.   (2,3)

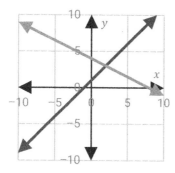

b.   (6,7)

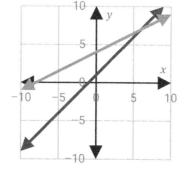

c.   (3,2)

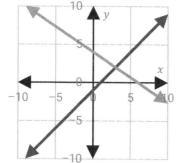

149

d. No solution

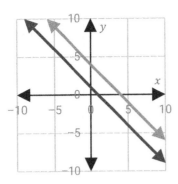

2. Solve the following system of equations using the graphing method: $\begin{cases} y = 2x + 3 \\ y = x + 5 \end{cases}$.

a. $(-2, 7)$

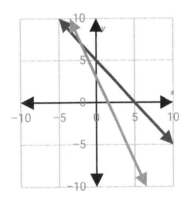

b. $(1, 5)$

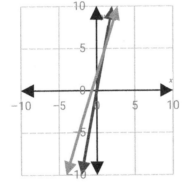

c. $(2, 7)$

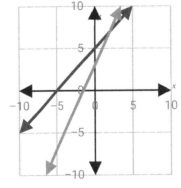

d. No solution

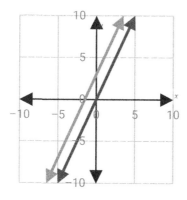

3. Solve the following system of equations using the graphing method: $\begin{cases} y = -2x + 4 \\ y = x - 2 \end{cases}$.

a. $(-2, 0)$

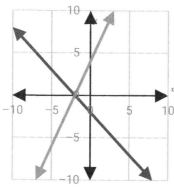

b. $(1, 2)$

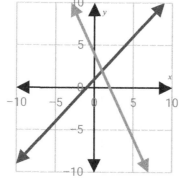

c. $(2, 0)$

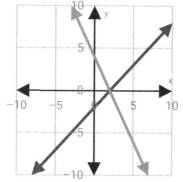

d. No solution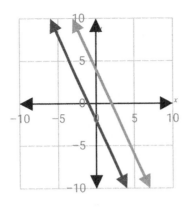

4. Solve the following system of equations using the graphing method: $\begin{cases} y = 2x - 1 \\ y = 3x \end{cases}$.

a. $(1, -3)$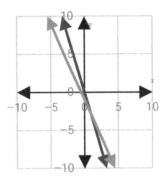

b. $(-1, -3)$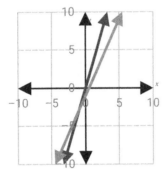

c. $(1, -3)$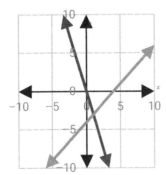

d. No solution

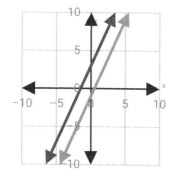

5. Solve the following system of equations using the graphing method: $\begin{cases} y = -3x + 5 \\ y = -4 \end{cases}$.

a. $(-4, -8)$

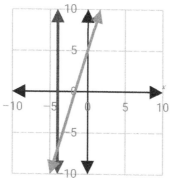

b. $(2, -2)$

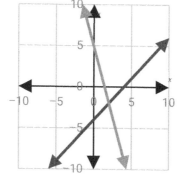

c. $(3, -4)$

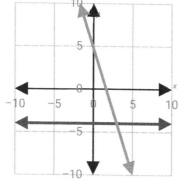

d. No solution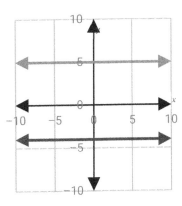

6. Solve the following system of equations using the graphing method: $\begin{cases} y = -\frac{2}{3}x - 2 \\ x = 3 \end{cases}$.

a. $(-4, -8)$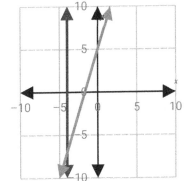

b. $(3, -4)$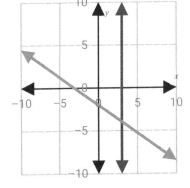

c. $(3, -4)$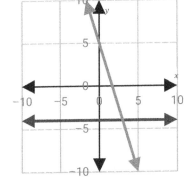

d. No solution

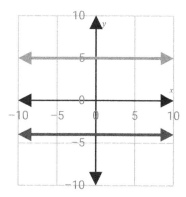

7. Solve the following system of equations using the graphing method: $\begin{cases} 3x + y = 5 \\ y = -3x - 2 \end{cases}$.

a. (1,1)

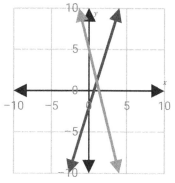

b. (0.5, −4)

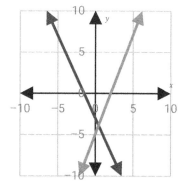

c. (1,2)

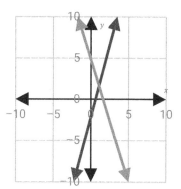

d.   No solution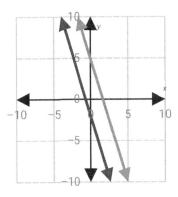

8. Solve the following system of equations using the graphing method: $\begin{cases} 2x - y = -1 \\ y = x - 1 \end{cases}$.

a. $(-2, -3)$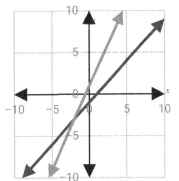

b. $(0, -1)$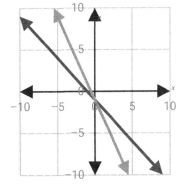

c. $(2, 3)$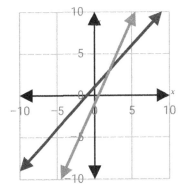

d. No solution

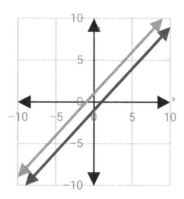

9. Solve the following system of equations using the graphing method: $\begin{cases} 2x - 3y = -3 \\ x = 3y - 9 \end{cases}$.

a. $(-6, 5)$

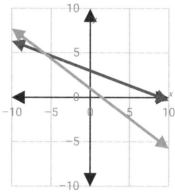

b. $(6, 5)$

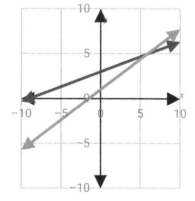

c. $(-4, 2)$

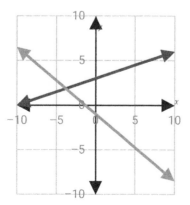

d. No solution

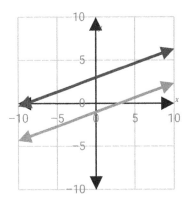

10. Solve the following system of equations using the graphing method: $\begin{cases} x + 4y = 12 \\ y = -\frac{1}{2}x + 1 \end{cases}$.

a. $(-2,0)$

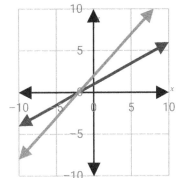

b. $(8,5)$

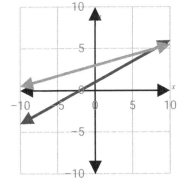

c. $(-8,5)$

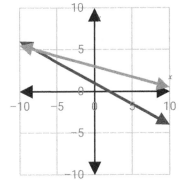

d. No solution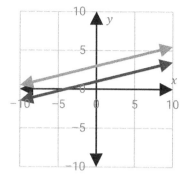

11. Solve the following system of equations using the graphing method: $\begin{cases} -4x - 2y = 6 \\ y = -\frac{4}{3}x - 5 \end{cases}$.

a. $(3, -9)$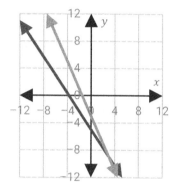

b. $(-3, -9)$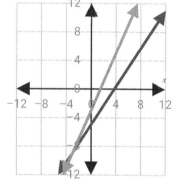

c. $(-0.5, -4)$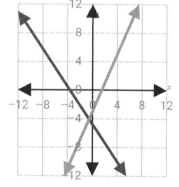

d. No solution

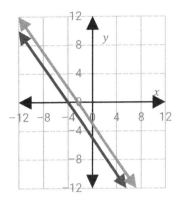

12. Solve the following system of equations using the graphing method: $\begin{cases} 6x - 3y = 9 \\ -4x + 2y = 14 \end{cases}$.

a. (2.5, 2)

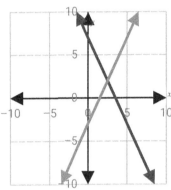

b. (−2.5, 2)

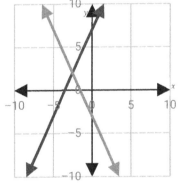

c. (−1, −5)

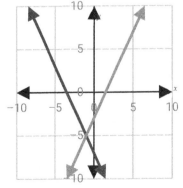

d. No solution

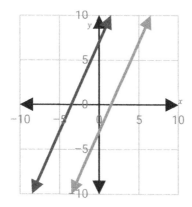

13. A store sells two types of picture frames. One frame sells for $3.00 and the other one sells for $4.00. The store sells 50 of the picture frames totaling $180 on a single day. Which of the following is the graph of the solution for the number of each frame sold where x represents the number of $3.00 frames sold and y represents the number of $4.00 frames sold?

a. $(-20, 30)$

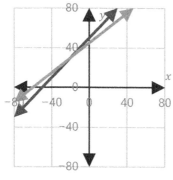

b. $(20, 30)$

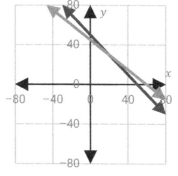

c. $(10, 40)$

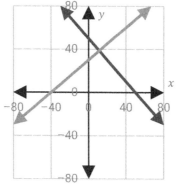

d. No solution

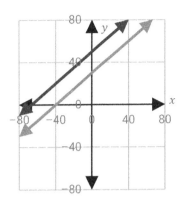

14. The sum of two numbers is −15. The one number is twice the other number. Which of the following is the graph for the solution of the two numbers?

a. (5, 10)

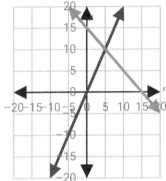

b. (5, −10)

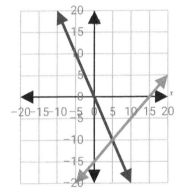

c. (−5, −10)

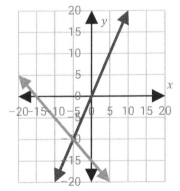

d. No solution

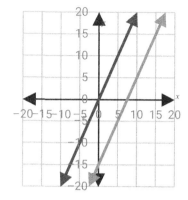

15. A paper bag contains a total of 20 nickels and dimes. The total amount of money in the bag is $1.40. Which of the following is the graph of the solution for the number of each coin in the bag where x is the number of nickels and y is the number of dimes?

a. $(-12, 8)$

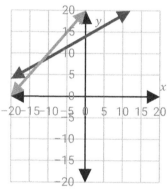

b. $(12, 8)$

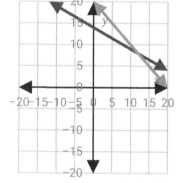

c. (8,12)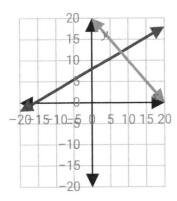

d. No solution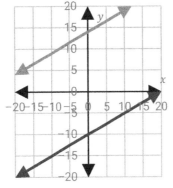

5.2 The Substitution Method

Practice Questions

1. Solve the system of equations using the substitution method: $\begin{cases} x = -5 \\ y = 2x + 8 \end{cases}$
    a. $(-1, 6)$
    b. $(1, 10)$
    c. $(-2, -5)$
    d. $(-5, -2)$

2. Solve the system of equations using the substitution method: $\begin{cases} y = 2x + 1 \\ y = x + 2 \end{cases}$
    a. $(3, 1)$
    b. $(1, 3)$
    c. $(2, 2)$
    d. $(-1, 1)$

3. Solve the system of equations using the substitution method: $\begin{cases} 3x + y = 10 \\ y = 2x \end{cases}$
    a. $(3, 4)$
    b. $(-1, -2)$
    c. $(2, 4)$
    d. $(4, 2)$

4. Use the substitution method to simplify the system of equations: $\begin{cases} y = -3x \\ 2x + 3y = -7 \end{cases}$
    a. $(2, -6)$
    b. $(7, -7)$
    c. $(3, -1)$
    d. $(1, -3)$

5. Use the substitution method to solve the system of equations: $\begin{cases} y = \frac{1}{2}x - 1 \\ y = x - 4 \end{cases}$
    a. $(-1, -5)$
    b. $(2, 0)$
    c. $(6, 2)$
    d. $(2, 6)$

6. Use the substitution method to solve the system of equations: $\begin{cases} y = x + 2 \\ x = \frac{1}{3}y + 4 \end{cases}$

   a. (7,9)
   b. (9,7)
   c. (6,6)
   d. (1,3)

7. Use the substitution method to solve the system of equations: $\begin{cases} x + 3y = 9 \\ 2x + y = -2 \end{cases}$

   a. (4,−3)
   b. (−3,4)
   c. (0,3)
   d. (−1,0)

8. Use the substitution method to solve the system of equations: $\begin{cases} 2x - y = 9 \\ x + 3y = 8 \end{cases}$

   a. (4,−1)
   b. (−1,3)
   c. (5,1)
   d. (1,5)

9. Use the substitution method to solve the system of equations: $\begin{cases} x + 4y = -6 \\ -2x + y = 3 \end{cases}$

   a. (2,1)
   b. (2,7)
   c. (−2,−1)
   d. (−1,−2)

10. Use the substitution method to solve the system of equations: $\begin{cases} -x - y = -12 \\ 2x + y = 16 \end{cases}$

    a. (8,4)
    b. (4,8)
    c. (−1,13)
    d. (−1,18)

11. Use the substitution method to solve the system of equations: $\begin{cases} x - 6y = -4 \\ 3x + 5y = 11 \end{cases}$

    a. (−3,4)
    b. (−10,1)
    c. (1,2)
    d. (2,1)

12. Use the substitution method to solve the system of equations: $\begin{cases} 2x + y = 10 \\ 4x + 2y = 6 \end{cases}$
   a. (2,6)
   b. (1,1)
   c. Infinitely many solutions
   d. No solution

13. For a school fundraiser, a science class sells chips and candy bars at a school's football game. The chips are sold for $1.50 each and the candy bars are sold for $1.25 each. The class raised a total of $127.50 from selling a total of 90 bags of chips and candy bars. How many of each item were sold?
   a. 60 bags of chips and 30 candy bars
   b. 30 bags of chips and 60 candy bars
   c. 45 bags of chips and 45 candy bars
   d. 70 bags of chips and 20 candy bars

14. The sum of two numbers is 20. Three times the smaller number equals the larger number. What are the two numbers?
   a. 6 and 14
   b. 5 and 15
   c. 4 and 12
   d. No solution

15. An acid solution is 10% acid. A second acid solution containing 50% acid is mixed with the first solution. How many liters of each solution should be mixed to have 60 liters of a 40% acid solution?
   a. 40 liters of the first solution and 20 liters of the second solution
   b. 30 liters of the first solution and 3 liters of the second solution
   c. 15 liters of the first solution and 45 liters of the second solution
   d. 45 liters of the first solution and 15 liters of the second solution

5.3 The Elimination Method

PRACTICE QUESTIONS

1. Use the elimination method to solve the system of equations: $\begin{cases} 2x + y = 7 \\ -2x + 3y = 5 \end{cases}$

   a. (2,3)
   b. (3,2)
   c. (1,5)
   d. (−1,1)

2. Use the elimination method to solve the system of equations: $\begin{cases} 2x + y = 13 \\ x + y = 8 \end{cases}$

   a. (4,4)
   b. (4,5)
   c. (5,3)
   d. (3,5)

3. Use the elimination method to solve the system of equations: $\begin{cases} x + 3y = 7 \\ x - y = -1 \end{cases}$

   a. (2,1)
   b. (1,2)
   c. (4,1)
   d. (5,6)

4. Use the elimination method to solve the system of equations: $\begin{cases} 2x - 4y = -4 \\ x + 3y = 13 \end{cases}$

   a. (6,12)
   b. (1,4)
   c. (3,4)
   d. (4,3)

5. Use the elimination method to solve the system of equations: $\begin{cases} 3x - y = -8 \\ 2x + 2y = 8 \end{cases}$

   a. (−2,2)
   b. (2,2)
   c. (−1,5)
   d. (5,−1)

6. Use the elimination method to solve the system of equations: $\begin{cases} x + 5y = -4 \\ -3x - 4y = 1 \end{cases}$

   a. (1,−1)
   b. (−1,1)
   c. (6,−2)
   d. (−3,2)

7. Use the elimination method to solve the system of equations: $\begin{cases} -5x - y = 7 \\ 2x + 3y = 5 \end{cases}$

   a. $(-1, -2)$
   b. $(1, 1)$
   c. $(3, -2)$
   d. $(-2, 3)$

8. Use the elimination method to solve the system of equations: $\begin{cases} 4x + 2y = 6 \\ -8x - 4y = -12 \end{cases}$

   a. $(1, 2)$
   b. $(2, 1)$
   c. No solution
   d. Infinite solutions

9. Use the elimination method to solve the system of equations: $\begin{cases} 2x - 3y = -6 \\ 3x - 2y = 1 \end{cases}$

   a. $(4, 3)$
   b. $(3, 4)$
   c. $(0, 2)$
   d. $(1, 2)$

10. Use the elimination method to solve the system of equations: $\begin{cases} x + 4y = 5 \\ 3x - 6y = 6 \end{cases}$

    a. $\left(3, \frac{1}{2}\right)$
    b. $\left(\frac{1}{2}, 3\right)$
    c. $(1, 1)$
    d. $(4, 1)$

11. Use the elimination method to solve the system of equations: $\begin{cases} y = -\frac{1}{5}x - \frac{3}{5} \\ -2x + 5y = 9 \end{cases}$

    a. $\left(5, \frac{2}{5}\right)$
    b. $\left(-5, -\frac{1}{5}\right)$
    c. $\left(-4, \frac{1}{5}\right)$
    d. $\left(\frac{1}{5}, -4\right)$

12. Use the elimination method to solve the system of equations: $\begin{cases} 6x - 3y = 12 \\ y = 2x + 5 \end{cases}$

   a. (3,2)
   b. (1,7)
   c. No solution
   d. Infinite solutions

13. The sum of two numbers, x and y is −60, and two times x minus y is −15. What are x and y?

   a. $x = -35$ and $y = -25$
   b. $x = -25$ and $y = -35$
   c. $x = -15$ and $y = -45$
   d. $x = -9$ and $y = -3$

14. For a fundraiser, your school sells bags of cotton candy and cans of soda at a school play. Cotton candy is sold for $2.25 a bag and sodas are sold for $2.00 each. Your school raised a total of $325 and sold a total of 150 bags of cotton candy and cans of soda. How many of each item were sold?

   a. 100 bags of cotton candy and 50 cans of soda
   b. 50 bags of cotton candy and 100 cans of soda
   c. 75 bags of cotton candy and 75 cans of soda
   d. 80 bags of cotton candy and 70 cans of soda

15. An acid solution is 20% acid. A second acid solution containing 60% acid is mixed with the first solution. How many liters of each solution should be mixed to have 90 liters of a 30% acid solution?

   a. 22.5 liters of the first solution and 67.5.5 liters of the second solution
   b. 67.5 liters of the first solution and 22.5 liters of the second solution
   c. 40 liters of the first solution and 50 liters of the second solution
   d. 75 liters of the first solution and 15 liters of the second solution

5.4 Writing Systems of Linear Equations

PRACTICE QUESTIONS

1. Write a system of two linear equations given a table of values for each equation. Then, state the solution to the system

Table 1

x	y
−2	−4
0	2
2	8
4	14

Table 2

x	y
−2	−6
0	−4
2	−2
4	0

a. $\begin{cases} y = 3x + 2 \\ y = x - 4 \end{cases}$ and the solution is $(-3, -7)$.

b. $\begin{cases} y = 3x - 2 \\ y = x + 4 \end{cases}$ and the solution is $(2, 4)$.

c. $\begin{cases} y = 3x \\ y = x + 2 \end{cases}$ and the solution is $(0, 0)$.

d. $\begin{cases} y = x - 4 \\ y = 3x \end{cases}$ and the solution is $(-3, -9)$.

2. Write a system of two linear equations given a table of values for each equation. Then, state the solution to the system.

Table 1

x	y
−3	−3
−1	1
1	5
5	13

Table 2

x	y
−3	−2
−1	0
1	2
5	6

a. $\begin{cases} y = 3x \\ y = x \end{cases}$ and the solution is $(0, 0)$.

b. $\begin{cases} y = 2x \\ y = 3x \end{cases}$ and the solution is $(3, 6)$.

c. $\begin{cases} y = 2x + 3 \\ y = x + 1 \end{cases}$ and the solution is $(-2, -1)$.

d. $\begin{cases} y = 3x - 2 \\ y = x - 1 \end{cases}$ and the solution is $(-1, -5)$.

3. Write a system of two linear equations given a table of values for each equation. Then, state the solution to the system.

Table 1

x	y
−3	−5
−1	−1
2	5
5	11

Table 2

x	y
−3	15
−1	9
2	0
5	−9

a. $\begin{cases} x - y = 4 \\ 3x + y = 6 \end{cases}$ and the solution is (6,4).

b. $\begin{cases} 2x - y = -2 \\ x + y = 5 \end{cases}$ and the solution is (2,3).

c. $\begin{cases} 2x + y = -1 \\ 3x - y = 6 \end{cases}$ and the solution is (1,−3).

d. $\begin{cases} 2x - y = -1 \\ 3x + y = 6 \end{cases}$ and the solution is (1,3).

4. Write a system of two linear equations given a table of values for each equation. Then, state the solution to the system.

Table 1

x	y
−2	10
1	7
2	6
4	4

Table 2

x	y
−2	−14
1	−8
2	−6
4	−2

a. $\begin{cases} x - y = -8 \\ 2x - y = 10 \end{cases}$ and the solution is (−6,−2).

b. $\begin{cases} x + y = 8 \\ 2x - y = 10 \end{cases}$ and the solution is (6,2).

c. $\begin{cases} x + y = 8 \\ x + 2y = -10 \end{cases}$ and the solution is (26,−18).

d. $\begin{cases} 3x + 4y = 17 \\ 2x - 4y = -10 \end{cases}$ and the solution is (15,10).

5. Write a system of two linear equations given a table of values for each equation. Then, state the solution to the system.

Table 1
x	y
−5	1
1	9
4	13
7	17

Table 2
x	y
−5	−7
1	17
4	29
7	41

a. $\begin{cases} 2x + y = 14 \\ y = 5x - 8 \end{cases}$ and the solution is (4,6).

b. $\begin{cases} 2x + y = 18 \\ y = 3x - 10 \end{cases}$ and the solution is (6,3).

c. $\begin{cases} 4x + 3y = 23 \\ y = 4x - 13 \end{cases}$ and the solution is (2,5).

d. $\begin{cases} 4x - 3y = -23 \\ y = 4x + 13 \end{cases}$ and the solution is (−2,5).

6. Write a system of two linear equations given a table of values for each equation. Then, state the solution to the system.

Table 1
x	y
−2	$-\frac{7}{5}$
−1	−1
1	$-\frac{1}{5}$
3	$\frac{3}{5}$

Table 2
x	y
−2	−3
−1	$-\frac{7}{3}$
1	−1
3	$\frac{1}{3}$

a. $\begin{cases} 2x - 5y = 3 \\ 2x - 3y = 5 \end{cases}$ and the solution is (4,1).

b. $\begin{cases} 2x + 5y = 3 \\ 2x - 3y = 5 \end{cases}$ and the solution is (1, −1).

c. $\begin{cases} x + y = -2 \\ 2x + 3y = -5 \end{cases}$ and the solution is (−1, −1).

d. $\begin{cases} 2x + 5y = 3 \\ x - y = -2 \end{cases}$ and the solution is (1,3).

7. Write a system of two linear equations given the graphs for each equation. Then state the solution to the system.

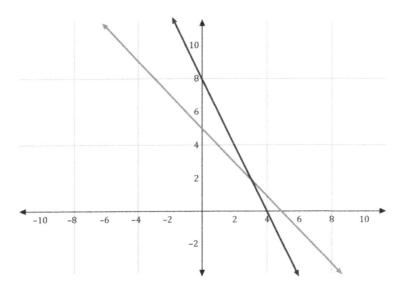

a. $\begin{cases} y = -x + 5 \\ y = -2x + 8 \end{cases}$ and the solution is $(3,2)$.

b. $\begin{cases} y = x + 5 \\ y = 2x + 8 \end{cases}$ and the solution is $(-3,2)$.

c. $\begin{cases} y = x - 5 \\ y = 2x - 8 \end{cases}$ and the solution is $(3,2)$.

d. $\begin{cases} y = -x - 5 \\ y = -2x + 4 \end{cases}$ and the solution is $(3,-2)$.

8. Write a system of two linear equations given the graphs for each equation. Then state the solution to the system.

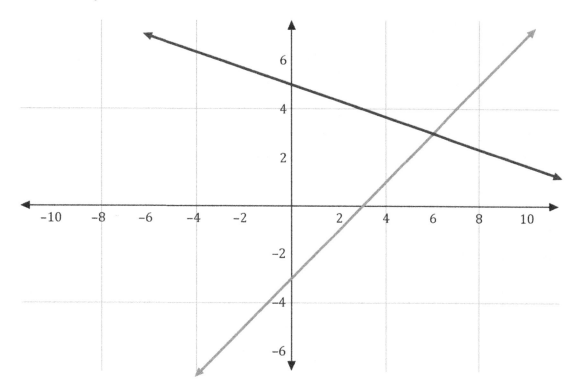

a. $\begin{cases} y = -x - 5 \\ y = \frac{1}{3}x - 5 \end{cases}$ and the solution is (6,3).

b. $\begin{cases} y = x - 3 \\ y = 3x + 5 \end{cases}$ and the solution is (3,6).

c. $\begin{cases} y = x - 3 \\ y = -\frac{1}{3}x + 5 \end{cases}$ and the solution is (6,3).

d. $\begin{cases} y = -x - 3 \\ y = \frac{1}{3}x + 5 \end{cases}$ and the solution is (-6,3).

9. Write a system of two linear equations given the graphs for each equation. Then state the solution to the system.

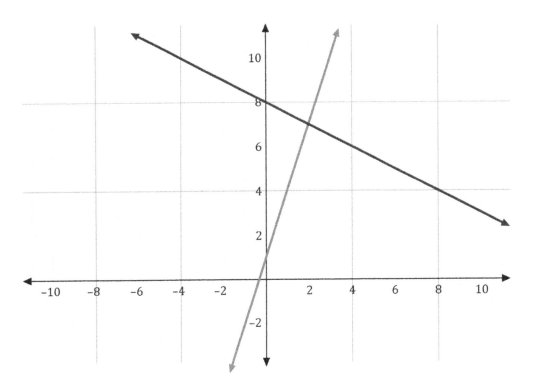

a. $\begin{cases} y = -3x + 1 \\ y = \frac{1}{2}x + 8 \end{cases}$ and the solution is $(-2,7)$.

b. $\begin{cases} y = 3x + 1 \\ y = -\frac{1}{2}x + 8 \end{cases}$ and the solution is $(2,7)$.

c. $\begin{cases} y = 3x + 1 \\ y = -2x - 8 \end{cases}$ and the solution is $(2,7)$.

d. $\begin{cases} y = \frac{1}{3}x - 1 \\ y = \frac{1}{2}x + 8 \end{cases}$ and the solution is $(-2,7)$.

10. Write a system of two linear equations given the graphs for each equation. Then state the solution to the system.

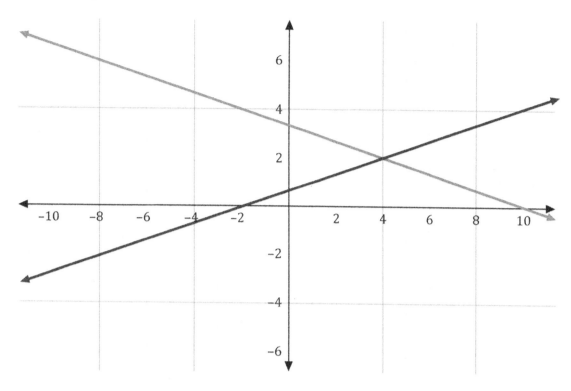

a. $\begin{cases} -x + 3y = 10 \\ x + 3y = 2 \end{cases}$ and the solution is $(-4, 2)$.

b. $\begin{cases} 3x + y = 30 \\ 3x - y = -6 \end{cases}$ and the solution is $(4, 2)$.

c. $\begin{cases} x - 3y = 10 \\ x + 3y = -2 \end{cases}$ and the solution is $(4, -2)$.

d. $\begin{cases} x + 3y = 10 \\ x - 3y = -2 \end{cases}$ and the solution is $(4, 2)$.

11. Write a system of two linear equations given the graphs for each equation. Then state the solution to the system.

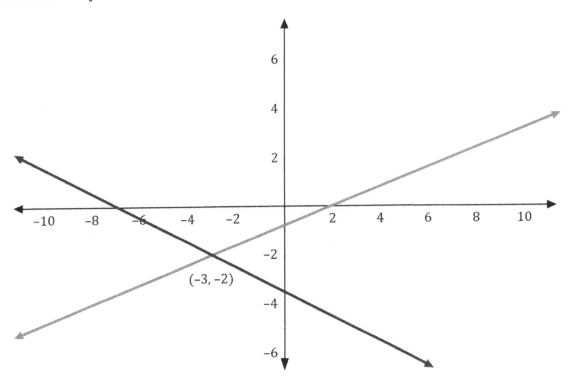

a. $\begin{cases} -2x + 5y = 4 \\ x + 3y = 6 \end{cases}$ and the solution is $(-3, -2)$.

b. $\begin{cases} x - y = 5 \\ x + 2y = -7 \end{cases}$ and the solution is $(3, -2)$.

c. $\begin{cases} 2x - 5y = 4 \\ x + 2y = -7 \end{cases}$ and the solution is $(-3, -2)$.

d. $\begin{cases} 2x + 5y = 4 \\ x - 2y = -7 \end{cases}$ and the solution is $(-3, 2)$.

12. Write a system of two linear equations given the graphs for each equation. Then state the solution to the system.

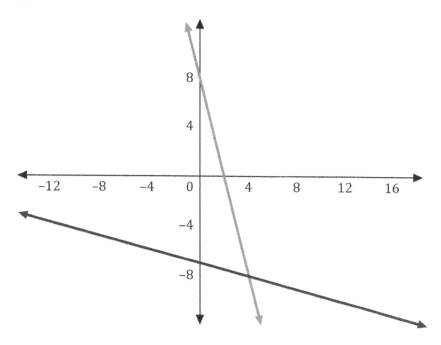

a. $\begin{cases} -4x + y = 8 \\ x - 4y = 28 \end{cases}$ and the solution is $(4, -8)$.

b. $\begin{cases} x + 4y = 8 \\ 4x + y = -28 \end{cases}$ and the solution is $(-8, 4)$.

c. $\begin{cases} 4x - y = 8 \\ x - 4y = -28 \end{cases}$ and the solution is $(-4, 8)$.

d. $\begin{cases} 4x + y = 8 \\ x + 4y = -28 \end{cases}$ and the solution is $(4, -8)$.

13. A small grocery store sold a total of 11 hygiene items one day—specifically packages of soap and bottles of shampoo. A package of soap sells for $3.00, and a bottle of shampoo sells for $8.00. The total sales for the soap and shampoo on that same day is $48.00. How many packages of soap and bottles of shampoo are sold that day?
    a. 18 packages of soap and 7 bottles of shampoo
    b. 8 packages of soap and 3 bottles of shampoo
    c. 3 packages of soap and 8 bottles of shampoo
    d. 24 packages of soap and 24 bottles of shampoo

14. A jar is filled with a total of 25 coins—specifically quarters and fifty-cent coins. The total value of the coins in the jar is $10.00. How many quarters and fifty-cent coins are in the jar?
    a. 12 quarters and 13 fifty-cent coins
    b. 15 quarters and 5 fifty-cent coins
    c. 10 quarters and 15 fifty-cent coins
    d. 10 quarters and 25 fifty-cent coins

15. The sum of two integers, x and y is –15, and x is 35 more than y. What are the values of x and y?
   a. $x = 10$ and $y = -25$
   b. $x = -10$ and $y = 25$
   c. $x = 10$ and $y = 25$
   d. $x = -10$ and $y = -25$

5.5 Graphing Solutions to Systems of Linear Inequalities

PRACTICE QUESTIONS

1. Graph the system $\begin{cases} x + y \geq 5 \\ 2x + y \leq 8 \end{cases}$, and determine if the point $(-2, 9)$ is in the solution set.

a. The given point is in the solution.

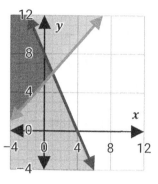

b. The given point is in the solution.

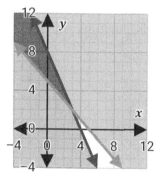

c. The given point is not in the solution.

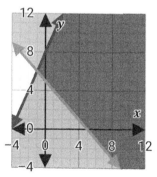

d. The given point is not in the solution.

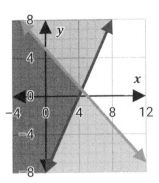

2. Graph the system $\begin{cases} 2x - y \leq 4 \\ x + 3y \geq 6 \end{cases}$, and determine if the point $(3, 5)$ is in the solution set.

a. The given point is in the solution.

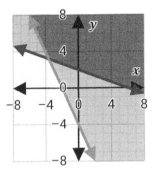

b. The given point is in the solution.

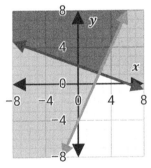

c. The given point is not in the solution.

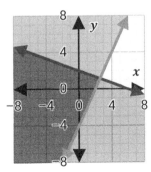

d. The given point is not in the solution.

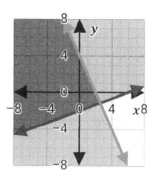

3. Graph the system $\begin{cases} 3x + y < 9 \\ x + 2y \geq 12 \end{cases}$, and determine if the point $(-3, 8)$ is in the solution set.

a. The given point is not in the solution.

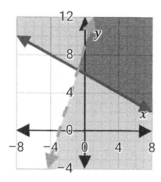

b. The given point is not in the solution.

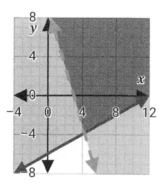

c. The given point is not in the solution.

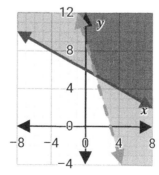

183

d. The given point is in the solution.

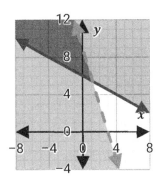

4. Graph the system $\begin{cases} x - 5y > 5 \\ 2x + 3y < 6 \end{cases}$, and determine if the point $(0, -3)$ is in the solution set.

a. The given point is in the solution.

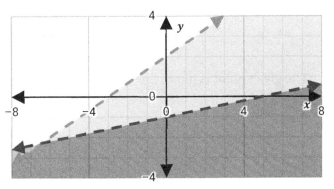

b. The given point is not in the solution.

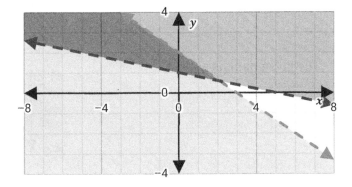

c. The given point is in the solution.

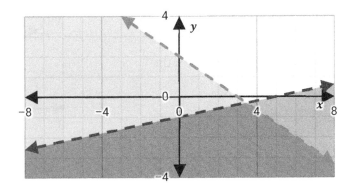

d. The given point is not in the solution.

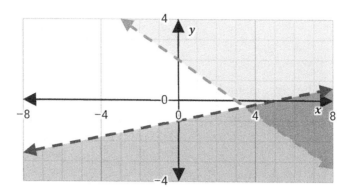

5. Graph the system $\begin{cases} 4x - y \leq 8 \\ x - y > 0 \end{cases}$, and determine if the point $(-2, 3)$ is included in the solution set.

a. The given point is in the solution.

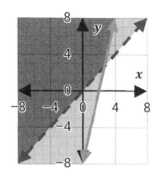

b. The given point is not in the solution.

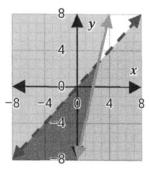

c. The given point is not in the solution.

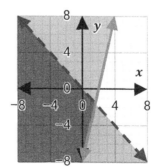

d. The given point is in the solution.

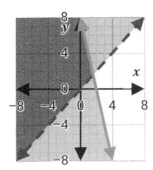

6. Graph the system $\begin{cases} 3x + 5y \geq 15 \\ 2x - 6y \leq 12 \end{cases}$, and determine if the point $(6, 2)$ is in the solution set.

a. The given point is in the solution.

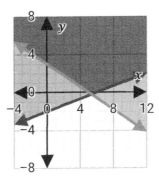

b. The given point is not in the solution.

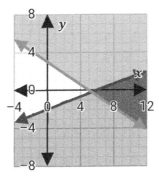

c. The given point is not in the solution.

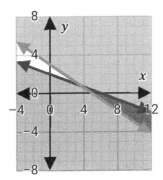

d. The given point is in the solution.

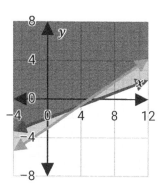

7. Graph the system $\begin{cases} y > 2x \\ x + y \geq -3 \end{cases}$, and determine if the point $(2, 4)$ is in the solution set.

a. The given point is not in the solution.

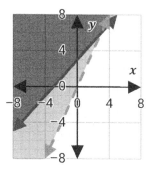

b. The given point is not in the solution.

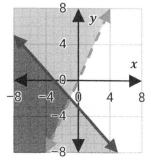

c. The given point is in the solution.

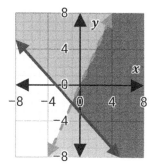

d. The given point is not in the solution.

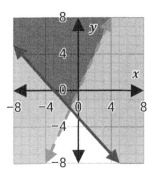

8. Graph the system $\begin{cases} 2x - y \geq 4 \\ y < \frac{2}{5}x \end{cases}$, and determine if the point $(2, 0)$ is in the solution set.

a. The given point is not in the solution.

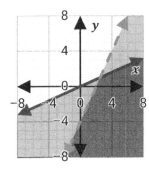

b. The given point is in the solution.

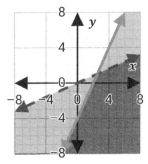

c. The given point is in the solution.

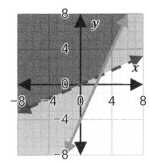

d. The given point is not in the solution.

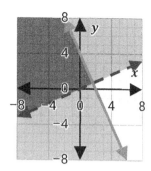

9. Graph the system $\begin{cases} y \leq 2x + 3 \\ y > x - 4 \end{cases}$, and determine if the point $(4, -3)$ is in the solution set.

a. The given point is not in the solution.

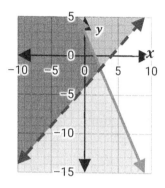

b. The given point is in the solution.

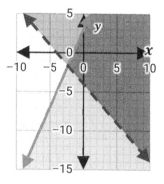

c. The given point is not in the solution.

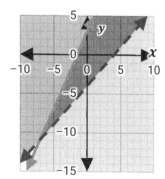

189

d. The given point is in the solution.

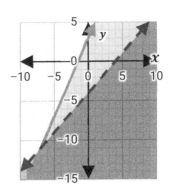

10. Graph the system $\begin{cases} y \leq -\frac{1}{3}x + 4 \\ y > \frac{3}{5}x \end{cases}$, and determine if the point $(0, 2)$ is in the solution set.

a. The given point is in the solution.

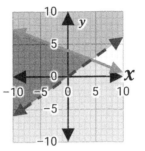

b. The given point is not in the solution.

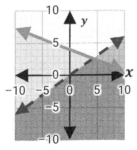

c. The given point is not in the solution.

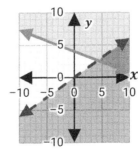

190

d. The given point is in the solution.

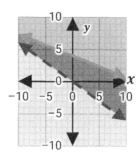

11. Graph the system $\begin{cases} y \leq -\frac{1}{2}x + 4 \\ y > \frac{3}{4}x \end{cases}$, and determine if the point $(6, 3)$ is in the solution set.

a. The given point is in the solution.

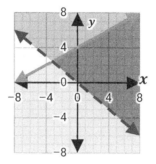

b. The given point is not in the solution.

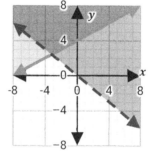

c. The given point is in the solution.

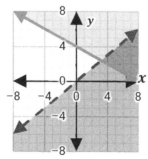

d. The given point is not in the solution.

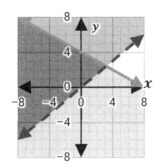

12. Graph the system $\begin{cases} y \leq 3 \\ x > -2 \end{cases}$, and determine if the point $(4, 2)$ is in the solution set.

a. The given point is in the solution.

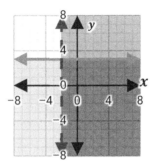

b. The given point is not in the solution.

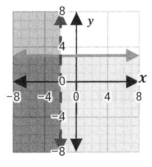

c. The given point is not in the solution.

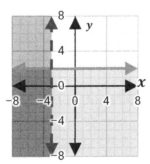

d. The given point is in the solution.

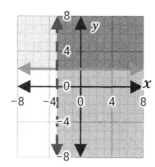

13. The sum of x and y is greater than or equal to twenty-five. The number y is greater than the sum of four and twice x. Which of the following graphs represents the possible values of x and y?

a.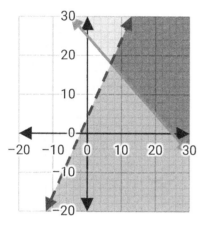

c.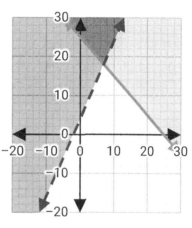

b.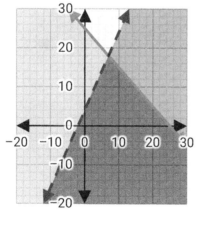

d.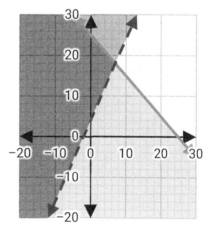

14. A department store never has more than thirty computer screens and keyboards on its shelf at any one time. If there are at least eight more keyboards than computer screens on the shelves, which of the following could be the graph of the shaded region for the various combinations of computer screens and keyboards the department store can have? Let x be the number of computer screens and y be the number of keyboards.

a.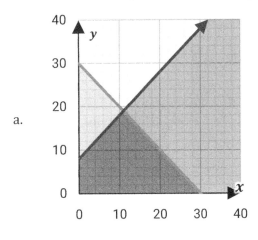

c.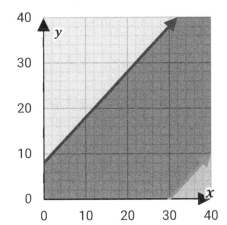

b.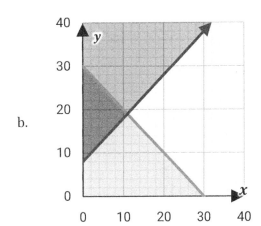

d.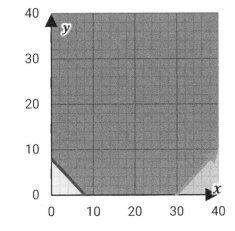

15. You have no more than fifteen hours to study for your upcoming tests in English and Biology. You spend less than twice as much time studying for your science test than your English test. If x is the number of hours spent studying for the English test, and y is the number of hours spent studying for the Biology test, which graph represents the region for the possible number of hours you can spend studying for the two tests?

a.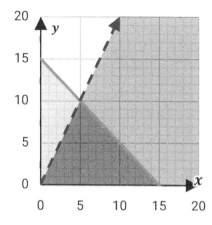

c.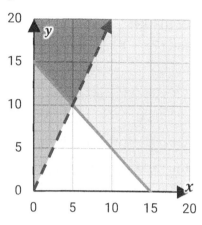

b.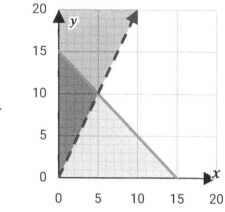

d.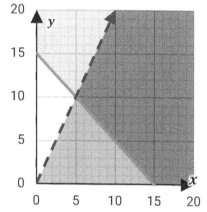

Chapter 6: Polynomials

6.1 Adding and Subtracting Polynomials

PRACTICE QUESTIONS

1. Find the sum: $(7x + 8) + (14x - 19)$.
    a. $21x^2 - 11$
    b. $21x + 27$
    c. $21x - 11$
    d. $10x$

2. Find the sum: $(14x^2 + 8x - 20) + (9x^2 - 10x + 32)$.
    a. $23x^2 - 2x + 12$
    b. $23x^4 - 2x^2 + 12$
    c. $21x^3 + 12$
    d. $23x^2 + 18x + 52$

3. Find the sum: $(11x + 15y - 9) + (7y - 24x - 35)$.
    a. $27y^2 - 57$
    b. $18x - 9y - 44$
    c. $9xy - 44$
    d. $-13x + 22y - 44$

4. Find the sum: $(5x^2 - 8y^2 - 14x - y + 16) + (x^2 + 3y^2 - 10x + 7y + 25)$.
    a. $-18x^2 + y^2 + 41$
    b. $6x^2 - 5y^2 - 24x + 6y + 41$
    c. $6x^2 - 24x + 42$
    d. $5x^2 - 5y^2 - 24x + 7y + 41$

5. Find the difference: $(-12x + 16) - (5x - 9)$.
    a. $-17x + 7$
    b. $-17x^2 + 25$
    c. $8x$
    d. $-17x + 25$

6. Find the difference: $(10x^2 + 14x + 18) - (9x^2 - x - 8)$.
    a. $x^2 + 14x + 26$
    b. $x^2 + 15x + 26$
    c. $x^2 + 13x + 10$
    d. $19x^2 + 13x + 10$

7. **Find the difference:** $(28x - 10y - 45) - (19x + 4y + 15)$.
   a. -27
   b. $9x - 14y - 60$
   c. $9x - 6y - 60$
   d. $9x - 6y - 30$

8. **Find the difference** $(4x^2 + 3xy - 2y^2 - 5x + 4y - 8) - (-2x^2 + 6xy + 7y^2 - 3x - 2y + 9)$.
   a. $6x^2 - 3xy - 9y^2 - 2x + 6y + 1$
   b. $6x^2 - 3xy - 9y^2 - 2x + 6y - 17$
   c. $6x^2 + 9xy + 5y^2 - 8x + 2y + 1$
   d. $2x^2 - 3xy - 9y^2 - 2x + 6y - 17$

9. **Find the sum:** $(2x^2 - 4x + 9) + (7x^2 - 9)$.
   a. $9x^2 - 4x - 18$
   b. $9x^2 - 4x$
   c. $5x^2$
   d. $9x^2 - 4x + 18$

10. **Find the difference:** $(-8x^2 + 5xy - 4y - 10) - (-8x^2 + 12y^2 + 5x - 6y - 10)$.
    a. $5xy + 12y^2 + 5x - 10y - 20$
    b. $5xy - 12y^2 - 5x + 2y - 20$
    c. $5xy - 12y^2 - 5x + 2y$
    d. $-16x^2 + 5xy - 12y^2 - 5x + 2y$

11. **Find the following sum:** $\left(\frac{1}{2}x + \frac{2}{9}y - \frac{2}{5}\right) + \left(\frac{1}{4}x + \frac{5}{9}y + \frac{3}{5}\right)$.
    a. $\frac{3}{2}x + \frac{7}{9}y + \frac{1}{5}$
    b. $\frac{3}{4}x + \frac{7}{9}y - 1$
    c. $\frac{3}{4}x + \frac{1}{3}y + \frac{1}{5}$
    d. $\frac{3}{4}x + \frac{7}{9}y + \frac{1}{5}$

12. **Find the difference** $(2.3x^2 - 4.9x + 7.8) - (4.1x^2 - 2.5x - 5.4)$.
    a. $-1.8x^2 - 2.4x + 13.2$
    b. $6.4x^2 - 2.4x + 13.2$
    c. $-1.8x^2 - 7.4x + 2.4$
    d. $-1.8x^2 - 2.4x - 2.4$

13. You have x nickels and 3 pennies in your left pocket. You have y dimes and 12 pennies in your right pocket. Represent the value, in cents, of the coins in each pocket as a polynomial, and then add the polynomials together to find an expression for the total value of coins you are carrying.
   a. $15x + 120y$ cents
   b. $5x + 10y + 15$ cents
   c. $8x + 22y$ cents
   d. $15xy + 15$ cents

14. You have a rectangular backyard with an area of $x^2 + 8x + 15$ square feet. In it you fence off a rectangular garden plot with an area of $x^2 - 6x + 8$ square feet. How much area is left in the backyard outside the garden plot?
   a. $14x + 7$ square feet
   b. $2x + 23$ square feet
   c. $-14x - 7$ square feet
   d. $2x^2 + 2x + 23$ square feet

15. You manufacture calculators that you sell for $30. If your factory's production rate is x calculators per day, then your daily production cost is $0.05x^2 + 12.45x + 500$ dollars. Assuming you sell every calculator that you produce, use polynomial subtraction to find an expression for your daily profit from producing x calculators per day.
   a. $0.05x^2 + 42.45x + 500$ dollars
   b. $17.5x - 500$ dollars
   c. $0.05x^2 - 17.55x + 500$ dollars
   d. $-0.05x^2 + 17.55x - 500$ dollars

6.2 Multiplying Monomials

PRACTICE QUESTIONS

1. Multiply $(3xy^2)(2x^4y)$.
    a. $6x^4y^2$
    b. $6x^5y^3$
    c. $5x^5y^3$
    d. $5x^2y^2$

2. Multiply $(4ab^3)(3a^5)$.
    a. $7a^6b^8$
    b. $7a^5b^{15}$
    c. $12a^6b^3$
    d. $12a^5b^3$

3. Multiply $(-5x^3y^3)(4x^2y^2)$.
    a. $-20x^5y^5$
    b. $-20x^6y^6$
    c. $-x^5y^5$
    d. x^9y^9

4. Multiply $(-4p^5q^7)(-6p^3q^4)$.
    a. $-24p^9q^{10}$
    b. $24p^{15}q^{28}$
    c. $24p^8q^{11}$
    d. $-24p^8q^{11}$

5. Multiply $\left(\frac{2}{3}x^5y^{-3}\right)(-12x^6y^5)$.
    a. $\frac{8x^{11}}{y^2}$
    b. $-\frac{8x^{30}}{y^{15}}$
    c. $8x^{11}y^2$
    d. $-8x^{11}y^2$

6. Multiply $\left(\frac{3}{5}r^2s^4\right)\left(\frac{1}{2}r^5s^{-1}\right)$.
    a. $\frac{3}{10}r^{10}s^3$
    b. $\frac{3}{10}r^7s^3$
    c. $\frac{2}{5}r^{10}s^3$
    d. $\frac{3r^{10}}{10s^3}$

7. Multiply $(2x^4y)^3(4x^2y^6)$.
   a. $32x^{14}y^9$
   b. $24x^{14}y^9$
   c. $32x^9y^{10}$
   d. $24x^6y^7$

8. Multiply $(-5x^3y)(-7x^2y^3)^2$.
   a. $-35x^5y^6$
   b. $-35x^5y^4$
   c. $-245x^7y^6$
   d. $-245x^7y^7$

9. Multiply $(-5p^3q^2)^2(p^2q^4)^3$.
   a. $25p^{10}q^9$
   b. $-25p^9q^{14}$
   c. $25p^{12}q^{16}$
   d. $-25p^{12}q^{16}$

10. Multiply $(4x^5y)^2 \left(\frac{3}{4}x^3y^4\right)^2$.
    a. $9x^{16}y^{10}$
    b. $6x^{16}y^{10}$
    c. $9x^{13}y^9$
    d. $6x^{16}y^{10}$

11. Multiply $\left(\frac{2}{5}p^3q\right)^2 (-5p^2q^2)^3 (p^4q^3)$.
    a. $20p^{16}q^{11}$
    b. $-20p^{16}q^{11}$
    c. $6p^{14}q^8$
    d. $-6p^9q^6$

12. Multiply $(7x^4y)^2 \left(\frac{2}{7}x^3y^4\right)^2 (-4x^2y)^2$.
    a. $-32x^9y^6$
    b. $32x^{16}y^{11}$
    c. $-64x^{18}y^{12}$
    d. $64x^{18}y^{12}$

13. Which of the following expressions represents the area of a rectangle that has a length of $3xy^2$ and a width of $2xy$?

    a. $5x^2y^2$
    b. $5x^2y^2$
    c. $6x^2y^3$
    d. $6xy^2$

14. Which of the following expressions represents the volume of a right circular cone that has a radius of $4mn^3$ and a height of $3m^2n$? Note: The formula for the volume, V, of a right circular cone is $V = \frac{\pi}{3}r^2h$.

    a. $16\pi m^3n^4$
    b. $16\pi m^4n^7$
    c. $4\pi m^4n^3$
    d. $4\pi m^3n^7$

15. Let z be the product of the two monomials a^2b^5 and $2a^4b$. Which of the following expressions represents z?

    a. $2a^6b^6$
    b. $2a^2b^5$
    c. $4a^6b^6$
    d. $4a^2b^5$

6.3 Dividing Monomials

PRACTICE QUESTIONS

1. Divide $15x^6$ by $5x^2$ and simplify.
   a. 9
   b. $3x^3$
   c. $3x^4$
   d. $10x^4$

2. Simplify the quotient $\frac{19x^3}{19x}$.
   a. 3
   b. $19x^2$
   c. x^3
   d. x^2

3. Simplify the quotient $(100z^8) \div (25z^7)$.
   a. $75z$
   b. $4z^{15}$
   c. $4z$
   d. $\frac{32}{7}$

4. Simplify the following quotient.
$$\frac{6x^8}{18x^2}$$
   a. $\frac{x^4}{3}$
   b. $3x^6$
   c. $\frac{4}{3}$
   d. $\frac{x^6}{3}$

5. Divide $14x^3$ by $7x^6$ and simplify.
   a. $\frac{2}{x^2}$
   b. $2x^3$
   c. $\frac{2}{x^3}$
   d. $2x^2$

6. Simplify the quotient $\frac{22x^8}{33x^{12}}$.

   a. $\frac{1}{11x^4}$
   b. $\frac{2}{3x^4}$
   c. $\frac{2x^2}{3x^3}$
   d. $\frac{4}{9}$

7. Simplify the quotient $(35x^{10}y^8) \div (5x^2y^2)$.

   a. $7x^5y^4$
   b. $7x^8y^6$
   c. 140
   d. $30x^8y^6$

8. Simplify the following quotient.

$$\frac{24x^9y^6}{36x^3y^6}$$

   a. $\frac{2x^3y}{3}$
   b. $\frac{2x^6y}{3}$
   c. $\frac{2x^3}{3}$
   d. $\frac{2x^6}{3}$

9. Divide $32s^2t^3$ by $8st$ and simplify.

   a. $4s^2t^3$
   b. $4st^2$
   c. $24st^2$
   d. 24

10. Simplify the quotient $-\frac{6xy^3}{15x^3y}$.

    a. $-\frac{2y^3}{5x^3}$
    b. $-\frac{y^2}{9x^2}$
    c. $-\frac{2y^2}{5x^2}$
    d. $-2x^4y^4$

11. Simplify the quotient $(54x^2y^3) \div (81x^4y^5)$.
    a. $2x^2y^2$
    b. $\frac{1}{27x^2y^2}$
    c. $\frac{1}{5}$
    d. $\frac{2}{3x^2y^2}$

12. Simplify the following quotient.
$$\frac{26x^{10}y^{12}}{13x^{20}y^4}$$
    a. 3
    b. $\frac{2y^8}{x^{10}}$
    c. $\frac{13y^8}{x^{10}}$
    d. $\frac{2y^3}{x^2}$

13. A rectangle has a length of $15y$ meters and an area of $90xy$ square meters. Use division of monomials to find the width of the rectangle.
    a. $6y$ meters
    b. $6x$ meters
    c. $1{,}350xy^2$ meters
    d. $\frac{1}{6x}$ meters

14. A rectangular solid box is $44y$ inches high and has volume $66xy^2$ cubic inches. Use division of monomials to find the area of the base of the box. Hint: The volume of a rectangular solid can be found by multiplying the area of the base by the height.
    a. $3x$ square inches
    b. $\frac{3xy^2}{2}$ square inches
    c. $\frac{2}{3xy}$ square inches
    d. $\frac{3xy}{2}$ square inches

15. You are a hot chocolate vendor. On a normal day, working in your food truck, you sell x cups of hot chocolate for y dollars each. But working at the concessions stand on game days at the local stadium, you triple the normal price and yet sell many more cups. On a particularly cold day working concessions at the stadium, you bring in $51xy$ dollars from selling hot chocolate. Use division of monomials to calculate how many cups you sold.
    a. $51y$ cups
    b. $17x$ cups
    c. $51x$ cups
    d. $17y$ cups

6.4 Multiplying Polynomials by Monomials

PRACTICE QUESTIONS

1. Simplify the product $\frac{1}{3}x(6x - 12)$.
    a. $2x - 12$
    b. $2x^2 - 12x$
    c. $2x^2 - 12$
    d. $2x^2 - 4x$

2. Simplify the product $11x(2x^2 + 3x + 5)$.
    a. $13x^3 + 14x^2 + 16x$
    b. $22x^3 + 33x^2 + 55x$
    c. $22x^2 + 88x$
    d. $22x^3 + 3x + 5$

3. Simplify the product $-4x^2(7x^2 - x - 9)$.
    a. $-28x^4 - 4x^3 - 36x^2$
    b. $-28x^4 + 4x^3 + 36x^2$
    c. $-28x^4 - x - 9$
    d. $-28x^4 + x^3 + 9x^2$

4. Simplify the product $2t^3(-6t^3 + 8t^2 - 3t + 4)$.
    a. $-12t^6 + 16t^5 - 6t^4 + 8t^3$
    b. $-12t^9 + 16t^6 - 6t^3 + 8$
    c. $-12t^6 + 8t^2 - 3t + 4$
    d. $-12t^6 + 8t^5 - 3t^4 + 4t^3$

5. Simplify the product $-3x^4(5x^6 - 7x^4 - 6x + 6)$.
    a. $-15x^{10} - 7x^8 - 6x^5 + 6x^4$
    b. $-15x^{10} + 21x^8 + 18x^5 - 18x^4$
    c. $-15x^{10} - 7x^4 - 6x + 6$
    d. $-15x^{10} - 21x^8 - 18x^5 + 18x^4$

6. Simplify the product $x^2y(2x^2 + 3x - 10)$.
    a. $2x^4y + 3x - 10$
    b. $-5x^2y$
    c. $2x^4 + 3x - 10y$
    d. $2x^4y + 3x^3y - 10x^2y$

7. Simplify the product $8x^5(-3x^2 - 2xy + 12y^2)$.
   a. $-24x^7 - 2xy + 12y^2$
   b. $-24x^7 - 2x^6y + 12x^5y^2$
   c. $-24x^{10} - 16x^5y + 96x^5y^2$
   d. $-24x^7 - 16x^6y + 96x^5y^2$

8. Simplify the product $\frac{3}{2}xy(6x^2 - 8xy + 3y^2)$.
   a. $9x^2y - 12xy + \frac{9}{2}xy^2$
   b. $9x^2y - 8xy + 3y^2$
   c. $\frac{15}{2}x^3y - 9.5x^2y^2 + \frac{9}{2}xy^3$
   d. $9x^3y - 12x^2y^2 + \frac{9}{2}xy^3$

9. Simplify the product $2st^2(27s^3 + 9s^2t + 3st^2 + t^3)$.
   a. $54s^4t^2 + 18s^3t^3 + 6s^2t^4 + 2st^5$
   b. $54s^4t^2 + 9s^3t^3 + 3s^2t^4 + st^5$
   c. $54s^3t^2 + 18s^2t^2 + 6st^2 + 2st^3$
   d. $54s^4t^2 + 9s^2t + 3st^2 + t^3$

10. Simplify the product $-10x^3y(4x^5 - 2y^5 + 9x^2y^2 - 3)$.
    a. $-40x^{15}y + 20x^3y^5 - 90x^6y^2 + 30x^3y$
    b. $-40x^8y + 20x^3y^6 - 90x^5y^3 + 30x^3y$
    c. $-40x^8y - 20x^3y^6 + 90x^5y^3 - 30x^3y$
    d. $-40x^8y - 2y^5 + 9x^2y^2 - 3$

11. Simplify the product $xy^2(-7x + 6y + 5xy)$.
    a. $4xy^2$
    b. $-7x^2y^2 + 6y + 5xy$
    c. $-7x^2y^2 + 6xy^3 + 5x^2y^3$
    d. $-7x^2 + 6y + 5xy^3$

12. Simplify the product $3x^2y^3(2x^2 + 6y^2 + 4x - 9)$.
    a. $6x^4y^3 + 18x^2y^5 + 12x^3y^3 - 27x^2y^3$
    b. $6x^4y^3 + 6x^2y^5 + 4y^3 - 9x^4y^3$
    c. $6x^4y^3 + 6y^2 + 4x - 9$
    d. $6x^4y^3 + 18x^2y^6 + 12x^2y^3 - 27x^2y^3$

13. Your company manufactures collectible plush animal toys. Last year you sold x toy penguins at a price of y dollars each. This year penguins have become wildly popular, and you expect to sell three times as many, even though you are going to raise the price by two dollars. Write an expression for the money you expect to receive from selling plush penguins next year in terms of x and y.

   a. $3xy + 2$ dollars
   b. $xy + 2x + 3y + 6$ dollars
   c. $3xy + 6x$ dollars
   d. $3xy + 6y$ dollars

14. You have a rectangular garden plot x feet long and y feet wide. You want to expand it by making it ten feet wider and four times as long. Write an expression for the area of the expanded plot in terms of x and y.

   a. $4xy + 40$ square feet
   b. $4xy + 40y$ square feet
   c. $xy + 10x + 4y + 40$ square feet
   d. $4xy + 40x$ square feet

15. You have a rectangular pen built out of fencing panels. Each panel is y meters long. The pen is twice as long as it is wide, with $2x$ panels along the length and x panels along the width. You are expanding the pen by adding a gate 2 meters long to each of the long sides of the rectangle. Write an expression for the area inside the expanded pen, measured in square meters.

   a. $2x^2y^2 + 2$ square meters
   b. $2x^2y^2 + 2xy$ square meters
   c. $4x^2y^2$ square meters
   d. $4xy + 4x + 4y + 4$ square meters

6.5 Multiplying Binomials

PRACTICE QUESTIONS

1. Multiply the following binomial: $(3x + 1)(x + 2)$.
    a. $3x^2 + 5x + 2$
    b. $3x^2 + 7x + 2$
    c. $3x^2 + 2$
    d. $3x^2 + 4$

2. Multiply the following binomial: $(4m - 1)(2m + 3)$.
    a. $8m^2 + 3$
    b. $8m^2 - 3$
    c. $8m^2 + 10m - 3$
    d. $8m^2 + 5m - 3$

3. Multiply the following binomials: $(4x + 9)(4x - 9)$.
    a. $16x^2 - 72x - 81$
    b. $16x^2 + 72x + 81$
    c. $16x^2 + 81$
    d. $16x^2 - 81$

4. Multiply the following binomials: $(3p - 2)(3p - 2)$.
    a. $9p^2 + 4$
    b. $9p^2 - 4$
    c. $9p^2 + 12p - 4$
    d. $9p^2 - 12p + 4$

5. Multiply the following binomials: $(-7r - 2)(3r + 4)$.
    a. $21r^2 - 22r + 8$
    b. $-21r^2 - 34r - 8$
    c. $12r^2 - 8$
    d. $-28r^2 - 34r - 6$

6. Multiply the following binomials: $(11m + 12p)(11m - 12p)$.
    a. $22m^2 - 264mp - 24p^2$
    b. $121m^2 - 46mp + 144p^2$
    c. $121m^2 - 144p^2$
    d. $22m^2 + 24p^2$

7. Multiply the following binomials: $(9x - 10y)(8x - 13y)$.
   a. $72x^2 - 40xy + 130y^2$
   b. $72x^2 - 197xy + 130y^2$
   c. $72x^2 + 130y^2$
   d. $72x^2 - 130y^2$

8. Multiply the following binomials: $(-9r + 4s)(-12r + 7s)$.
   a. $-3r^2 - 111rs + 28s^2$
   b. $108r^2 - 10rs + 11s^2$
   c. $108r^2 - 111rs + 28s^2$
   d. $-108r^2 + 111rs + 28s^2$

9. Multiply the following binomials: $(18x + 13y)(18x - 13y)$.
   a. $36x^2 - 62xy + 26y^2$
   b. $36x^2 + 72xy + 26y^2$
   c. $324x^2 + 169y^2$
   d. $324x^2 - 169y^2$

10. Multiply the following binomials: $(8c + 12d)(8c + 12d)$.
   a. $96c^2 + 192cd + 96d^2$
   b. $16c^2 + 40cd + 24d^2$
   c. $64c^2 + 40cd + 144d^2$
   d. $64c^2 + 192cd + 144d^2$

11. Multiply the following binomials: $(4m - 10n)(5m + 10n)$.
   a. $20m^2 + 10mn - 100n^2$
   b. $20m^2 - 10mn - 100n^2$
   c. $9m^2 - 10mn - 120n^2$
   d. $10m^2 + 10nm - 50n^2$

12. Multiply the following binomials: $(12x + 18y)(12x - 18y)$.
   a. $144x^2 - 324y^2$
   b. $144x^2 + 324y^2$
   c. $144x^2 + 432xy + 324y^2$
   d. $144x^2 + 60xy - 324y^2$

13. Which of the following expressions represents the area of a rectangle that has a length of $4m + 2$ and a height of $7m - 3$?
   a. $28m^2 + 6$
   b. $28m^2 - 6$
   c. $28m^2 + 2m - 6$
   d. $28m^2 - 2m - 6$

14. A square garden is bordered by a walkway of uniform width. The sides of the garden are x feet in length and the walkway is y feet wide. What is the total area of the garden and the walkway that surrounds it?

   a. $x^2 + 4xy + y^2$
   b. $x^2 + 2xy + 4y^2$
   c. $x^2 + 4xy + 4y^2$
   d. $x^2 + 4y^2$

15. Let $(3p + 5q)$ and $(4p - 9q)$ represent two integers. Which of the following represents their product?

   a. $12p^2 - 45p^2$
   b. $12p^2 - 45p^2$
   c. $12p^2 + 7pq - 45p^2$
   d. $12p^2 - 7pq - 45p^2$

6.6 Dividing Polynomials by Monomials

PRACTICE QUESTIONS

1. Divide $x^3 - 3x^2 + 6x$ by x.
    a. $x^2 - x + 2$
    b. $x^2 - 3x + 6$
    c. $x^4 - 3x^3 + 6x^2$
    d. $x^5 - x^4 + 2x^2$

2. Divide $4y^3 + y^2 - 8y$ by y.
    a. $y^3 + y^2 - 2y^2$
    b. $4y^4 + y^3 - 8y^2$
    c. $y^2 + y - 2$
    d. $4y^2 + y - 8$

3. Divide $3m^4 - 6m^3 + 9m^2$ by $3m$.
    a. $m^3 - 2m^2 + 3m$
    b. $m^3 - 3m^2 + 6m$
    c. $m^4 - 2m^3 + 3m^2$
    d. $m^4 - 3m^3 + 6m^2$

4. Divide $2a^4 + 5a^3 - 6a^2$ by a^2.
    a. $2a^6 + 5a^5 - 6a^4$
    b. $2a^2 + 5a - 6$
    c. $2a^8 + 5a^6 - 6a^4$
    d. $2a^2 + 5a^3 - 6$

5. Divide $10x^5 - 25x^3 + 5x^2$ by $5x^2$.
    a. $5x^7 - 20x^5 + x^4$
    b. $2x^{10} - 5x^6 + x^4$
    c. $2x^3 - 5x + 1$
    d. $5x^3 - 20x + 1$

6. Divide $4x^4y^3 + 6x^3y^2 + 2xy$ by $2xy$.
    a. $2x^3y^2 + 3x^2y + 1$
    b. $2x^3y^2 + 4x^2y + 1$
    c. $2x^5y^4 + 3x^4y^3 + x^2y^2$
    d. $2x^4y^3 + 3x^3y^2 + xy$

7. Divide $8x^2y^3 - 16x^3y - 24x^3y^2$ by $-4x^2y$.
   a. $2x^4y^3 - 4x^6y^2 - 6x^6y^2$
   b. $-2x^4y^4 + 6x^5y^2 + 4x^5y^3$
   c. $-2y^2 + 6xy - 4x$
   d. $-2y^2 + 6xy + 4x$

8. Divide $m^3n^4 - 9m^4n^3 + 6m^2n^2 - 15mn^2$ by $3mn^2$.
   a. $\frac{1}{3}m^2n^2 - 3m^3n + 2m - 5$
   b. $\frac{1}{3}m^3n^8 - 3m^4n^6 + 2m^2n^4 - 5m^2n^4$
   c. $-2m^3n^8 - 12m^4n^6 + 3m^2n^4 - 18m^2n^4$
   d. $-2m^2n^2 - 12m^3n + 3m - 18$

9. Divide $3p^5q^2 + 6p^6q^3 + 3p^2q^3 + 12p^2q^2$ by $3p^2q$.
   a. $p^2q + 23q^2 + pq^2 + 9pq^3$
   b. $p^7q^3 + 3p^8q^4 + p^4q^3 + 9p^4q^3$
   c. $p^3q + 2p^4q^2 + q^2 + 4q$
   d. $p^2q + 2p^2q^2 + pq^2 + 4pq^3$

10. Divide $-x^7y^5 + 9x^4y^4 - 36x^3y^3$ by $-9x^3y^2$.
    a. $-8x^4y^3 - 18xy^2 - 27y$
    b. $\frac{1}{9}x^4y^3 - xy^2 + 4y$
    c. $-2y^2 - 6xy - 4x$
    d. $\frac{1}{9}x^{10}y^7 - x^7y^6 + 4x^6y^5$

11. Divide $m^{11}n^8 - 4m^8n^6 + 5m^6n^4$ by m^5n^3.
    a. $m^{16}n^{11} - 3m^{13}n^9 + 4m^{11}n^7$
    b. $m^6n^5 - 3m^3n^3 + 4mn$
    c. $m^6n^5 - 4m^3n^3 + 5mn$
    d. $m^{16}n^{11} - 4m^{13}n^9 + 5m^{11}n^7$

12. Divide $-5x^9y^5 + 10x^7y^3 - 15x^6y^3 + 20x^5y$ by $-10x^5y$.
    a. $\frac{1}{2}x^{14}y^5 - x^{12}y^4 + \frac{3}{2}x^{11}y^4 - 2x^{10}y^2$
    b. $-15x^{14}y^5 - x^{12}y^4 + 5x^{11}y^4 - 10x^{10}y^2$
    c. $\frac{1}{2}x^4y^4 - x^2y^2 - \frac{3}{2}xy^2 - 2$
    d. $\frac{1}{2}x^4y^4 - x^2y^2 + \frac{3}{2}xy^2 - 2$

13. The area of a rectangular swimming pool is given by the polynomial $3n^4 + 6n^3 + n^2$. If the width of the pool is $3n^2$, what is the length of the pool?

   a. $n^2 + 2n + \frac{1}{3}$
   b. $n^6 + 2n^5 + \frac{1}{3}n^4$
   c. $n^6 + 3n^5 - 2n^4$
   d. $n^2 + 3n - 2$

14. The revenue, in dollars, generated by selling computer laptop batteries at a price of $3xy$ dollars each can be represented by the expression $24xy^2 + 9x^3y - 12xy$. If the revenue of the batteries can be calculated by the product of the price and the number of units sold, what is the number of batteries sold in terms of x and y?

   a. $21x^2y^3 + 6x^4y^2 - 9x^2y^2$
   b. $21x^2 + 6y - 9$
   c. $3x^2 + 8y - 4$
   d. $3x^4y^2 + 8x^2y^3 - 4x^2y^2$

15. The volume of a rectangular box is $2m^2n^3 + 8mn^2 + 10m^2n$. If the product of the length and width is $2mn$, what is the height of the box in terms of m and n?

   a. $m^3n^4 + 6m^2n^3 + 8m^3n^2$
   b. $mn^2 + 6n + 8m$
   c. $m^3n^4 + 4m^2n^3 + 5m^3n^2$
   d. $mn^2 + 4n + 5m$

6.7 Dividing Polynomials by Binomials

PRACTICE QUESTIONS

1. Use polynomial long division to calculate the quotient $(4x^2 + 27x + 35) \div (x + 5)$.
   a. $4x + 47$
   b. $4x + 7 + \frac{70}{x+5}$
   c. $4x + 7$
   d. $4x + 47 + \frac{87}{x+5}$

2. Use polynomial long division to calculate the quotient $(6x^2 + 53x + 59) \div (x + 8)$.
   a. $6x + 101 + \frac{747}{x+8}$
   b. $6x + 101 + \frac{867}{x+8}$
   c. $6x + 5 + \frac{19}{x+8}$
   d. $6x + 5 + \frac{99}{x+8}$

3. Use polynomial long division to calculate the quotient $(-3x^2 + 8x + 9) \div (x + 4)$.
   a. $-3x - 4 + \frac{25}{x+4}$
   b. $-3x + 20 - \frac{71}{x+4}$
   c. $-3x + 20 + \frac{89}{x+4}$
   d. $-3x - 4$

4. Use polynomial long division to calculate the quotient $(10x^2 + 3x - 13) \div (x - 1)$.
   a. $10x + 13$
   b. $10x + 13 - \frac{26}{x-1}$
   c. $10x - 7$
   d. $10x - 7 - \frac{20}{x-1}$

5. Use polynomial long division to calculate the quotient $(-6x^2 + 41x - 29) \div (x - 6)$.
   a. $-6x + 77 - \frac{491}{x-6}$
   b. $-6x + 5 - \frac{59}{x-6}$
   c. $-6x + 5 + \frac{1}{x-6}$
   d. $-6x + 77 + \frac{433}{x-6}$

6. Use polynomial long division to calculate the quotient $(5x^2 + 58x - 29) \div (x + 12)$.
  a. $5x + 118$
  b. $5x + 118 + \frac{1,387}{x+12}$
  c. $5x - 2 - \frac{53}{x+12}$
  d. $5x - 2 - \frac{5}{x+12}$

7. Use polynomial long division to calculate the quotient $(11x^2 - 120x + 189) \div (x - 9)$.
  a. $11x - 21$
  b. $11x - 219 + \frac{2,160}{x-9}$
  c. $11x - 219$
  d. $11x - 21 + \frac{378}{x-9}$

8. Use polynomial long division to calculate the quotient $(8x^2 - 19x + 39) \div (x - 2)$.
  a. $8x - 35 + \frac{109}{x-2}$
  b. $8x - 35 - \frac{31}{x-2}$
  c. $8x - 3 + \frac{45}{x-2}$
  d. $8x - 3 + \frac{33}{x-2}$

9. Use polynomial long division to calculate the quotient $(6x^2 - 23x + 3) \div (2x - 5)$.
  a. $3x - 19 - \frac{92}{2x-5}$
  b. $3x - 4 - \frac{17}{2x-5}$
  c. $3x - 19 + \frac{98}{2x-5}$
  d. $3x - 4 + \frac{23}{2x-5}$

10. Use polynomial long division to calculate the quotient $(3x^2 - 5x - 38) \div (x + 3)$.
  a. $3x - 14 - \frac{80}{x+3}$
  b. $3x + 4 - \frac{50}{x+3}$
  c. $3x - 14 + \frac{4}{x+3}$
  d. $3x + 4 - \frac{26}{x+3}$

11. Use polynomial long division to calculate the quotient $(7x^2 - 61x - 1,987) \div (x - 13)$.
    a. $7x - 152 - \frac{11}{x-13}$
    b. $7x - 152$
    c. $7x + 30 - \frac{1,597}{x-13}$
    d. $7x + 30 - \frac{2,377}{x-13}$

12. Use polynomial long division to calculate the quotient $(12x^2 - x - 88) \div (3x + 8)$.
    a. $4x + 11 + \frac{176}{3x+8}$
    b. $4x - 11 - \frac{176}{3x+8}$
    c. $4x - 11$
    d. $4x + 11$

13. A rectangle has an area of $4x^2 + 29x + 50$ square centimeters. Its width is $x + 5$ centimeters. Use polynomial long division to find the length of the rectangle.
    a. $4x + 10$ centimeters
    b. $4x + 12 - \frac{10}{x+5}$ centimeters
    c. $4x + 11 - \frac{5}{x+5}$ centimeters
    d. $4x + 9 + \frac{5}{x+5}$ centimeters

14. A business has a square parking lot with an area of x^2 square meters. Because business is booming, the owner wants to expand it into a rectangular lot 10 meters wider than the current lot with a new area of $2x^2 + 19x - 10$ square meters. Use polynomial long division to find the length of the new lot.
    a. $2x - 2$ meters
    b. $2x + 2$ meters
    c. $2x + 1$ meters
    d. $2x - 1$ meters

15. You manufacture widgets. Each one costs you $12 to make. As you raise the selling price, your profit per widget increases, but you sell fewer widgets per week. When you set the selling price at x dollars, your total profit for all widget sales during the week is $-3x^2 + 536x - 6,000$ dollars. Use polynomial long division to find the number of widgets you sell per week.
    a. $400 - 3x$ widgets
    b. $500 - 3x$ widgets
    c. $600 - 3x$ widgets
    d. $750 - 3x$ widgets

6.8 Factoring Using Common Monomial Factors

PRACTICE QUESTIONS

1. Factor the polynomial $12x + 18$ by factoring out the greatest common factor.
    a. $3(4x + 6)$
    b. $6(2x + 3)$
    c. $6x(2x + 3)$
    d. $12\left(x + \frac{3}{2}\right)$

2. Factor the polynomial $15x^4 + 12x^2$ by factoring out the greatest common factor.
    a. $3x^2(5x^2 + 4)$
    b. $5x^2(3x^2 + 12x)$
    c. $3x^2(15x^2 + 12)$
    d. $3(5x^2 + 4)$

3. Factor the polynomial $10x^3 - 20x^2 - 35x$ by factoring out the greatest common factor.
    a. $10(2x^2 - 4x - 7)$
    b. $5x(2x^2 - 4x - 7)$
    c. $x(10x^2 - 20x - 35)$
    d. $5x(x^2 - 2x - 3.5)$

4. Completely factor the polynomial $-6x^5 + 22x^4 - 18x^3$ by factoring out the greatest common factor.
    a. $-2x^3(3x^2 - 11x + 9)$
    b. $x^3(-6x^2 + 22x - 18)$
    c. $-2x^2(3x^3 - 11x^2 + 9x)$
    d. $-6x^3(x^2 - 4x + 3)$

5. Factor the polynomial $27x^{12} + 72x^8 - 36x^4$ by factoring out the greatest common factor.
    a. $9x^4(3x^8 + 8x^4 - 4)$
    b. $9x^4(3x^3 + 8x^2 - 4x)$
    c. $3x^4(9x^8 + 24x^4 - 12)$
    d. $12x^4(2x^8 + 6x^4 - 3)$

6. Factor the polynomial $6x^4 + 15x^3 + 10x^2$ by factoring out the greatest common factor.
    a. $x^2(6x^2 + 15x + 10)$
    b. $3x^2(2x^2 + 5x + 3)$
    c. $5x^2(x^2 + 3x + 2)$
    d. $2x^2(3x^2 + 7x + 5)$

7. Factor the polynomial $8x^2y - 12xy^2$ by factoring out the greatest common factor.
   a. $4x(2xy - 3y^2)$
   b. $xy(8x - 12y)$
   c. $4y(2x^2 - 3xy)$
   d. $4xy(2x - 3y)$

8. Completely factor the polynomial $2x^2y^3 + 26y^5$ by factoring out the greatest common factor.
   a. $2x^2(y^3 + 13y^5)$
   b. $y^3(2x^2 + 26y^2)$
   c. $2y^3(x^2 + 13y^2)$
   d. $2y^2(x^2y + 13y^3)$

9. Factor the polynomial $30x^6y^3 - 40x^5y^4 + 10x^4y^5$ by factoring out the greatest common factor.
   a. $5x^4y^3(6x^2 - 8xy + 2y^2)$
   b. $10x^4y^3(3x^2 - 4xy + y^2)$
   c. $30x^4y^3(x^2 - 40xy + 10y^2)$
   d. $xy(30x^5y^2 - 40x^4y^3 + 10x^3y^4)$

10. Completely factor the polynomial $-15x^6y^2 + 9x^2y^3 - 12x^3y^4$ by factoring out the greatest common factor.
    a. $-3x^2y^2(5x^4 - 3y + 4xy^2)$
    b. $-x^2y^2(15x^4 - 9y + 12xy^2)$
    c. $-3xy(5x^5y - 3xy^2 + 4x^2y^3)$
    d. $-15x^2y^2(x^4 - 9y + 12xy^2)$

11. Factor the polynomial $10x^2y^3 + 20xy^4 - 15x^2y^4 + 40xy^5$ by factoring out the greatest common factor.
    a. $xy^3(10x + 20y - 15xy + 40y^2)$
    b. $5xy^3(2x + 4y - 3xy + 8y^2)$
    c. $10xy(xy^2 + 20y^3 - 15xy^3 + 40y^4)$
    d. $5xy(2xy^2 + 4y^3 - 3xy^3 + 8y^4)$

12. Factor the polynomial $54z^5w - 36z^3w^3 - 24z^2w^5$ by factoring out the greatest common factor.
    a. $2z^2w(27z^3 - 18zw^2 - 12w^4)$
    b. $3z^2w(18z^3 - 12zw^2 - 8w^4)$
    c. $12z^2w(5z^3 - 3zw^2 - 2w^4)$
    d. $6z^2w(9z^3 - 6zw^2 - 4w^4)$

13. A rectangle has length $5x$ cm and area $15x^2 + 20x$ cm^2. Factor the length out of the area to rewrite the area in the form Area = length × width.
    a. Area = $5x(7x)$ cm^2
    b. Area = $5x(3x + 4)$ cm^2
    c. Area = $5x(3x + 20)$ cm^2
    d. Area = $5x(23x)$ cm^2

14. A rectangular solid (a box) has a square base with an area of x^2 ft^2 and a volume of $2x^3 + x^2y^2$ ft^3. Factor the area of the base out of the volume to rewrite the volume in the form Volume = (area of base) × (height).
    a. Volume = $x^2(2x + y^2)$ ft^3
    b. Volume = $x^2(2x^3 + y^2)$ ft^3
    c. Volume = $x^2(2x + x^2y^2)$ ft^3
    d. Volume = $x^2(1 + y^2)$ ft^3

15. You sell widgets for a price of $3x$ dollars per widget. Last month you had a total income of $3x^2 + 3xy + 6x^2y^2$ dollars from widget sales. Factor the price out of the total income to rewrite your total income in the form: Total Income = (price per item) × (number of items sold).
    a. Total Income = $3x(x + 3xy + 6x^2y^2)$ dollars
    b. Total Income = $3x(x + y + 2xy^2)$ dollars
    c. Total Income = $3x(x + 3y + 6xy^2)$ dollars
    d. Total Income = $3x(x^2 + y + 2x^2y^2)$ dollars

6.9 Factoring Trinomials of the Form

PRACTICE QUESTIONS

1. Factor the trinomial $x^2 + 12x + 11$ into a product of binomials.
    a. $(x + 1)(x + 12)$
    b. $(x + 1)(x + 10)$
    c. $(x + 1)(x + 11)$
    d. $(x + 5)(x + 6)$

2. Factor the trinomial $x^2 + 10x + 24$ into a product of binomials.
    a. $(x + 4)(x + 6)$
    b. $(x + 1)(x + 24)$
    c. $(x + 2)(x + 12)$
    d. $(x + 3)(x + 8)$

3. Factor the trinomial $x^2 + 24x + 144$ into a product of binomials.
    a. $(x + 6)(x + 24)$
    b. $(x + 9)(x + 16)$
    c. $(x + 8)(x + 18)$
    d. $(x + 12)^2$

4. Factor the trinomial $x^2 - 14x + 24$ into a product of binomials.
    a. $(x - 3)(x - 8)$
    b. $(x - 2)(x - 12)$
    c. $(x - 1)(x - 24)$
    d. $(x - 4)(x - 6)$

5. Factor the trinomial $x^2 - 18x + 45$ into a product of binomials.
    a. $(x - 3)(x - 15)$
    b. $(x - 1)(x - 45)$
    c. $(x - 5)(x - 9)$
    d. $(x - 9)^2$

6. Factor the trinomial $x^2 - 10x + 25$ into a product of binomials.
    a. $(x - 4)(x - 6)$
    b. $(x - 1)(x - 25)$
    c. $(x + 5)(x - 5)$
    d. $(x - 5)^2$

7. Factor the trinomial $x^2 + 5x - 24$ into a product of binomials.
    a. $(x + 3)(x - 8)$
    b. $(x + 9)(x - 4)$
    c. $(x + 6)(x - 4)$
    d. $(x + 8)(x - 3)$

8. Factor the trinomial $x^2 + 3x - 28$ into a product of binomials.
   a. $(x + 14)(x - 2)$
   b. $(x + 4)(x - 7)$
   c. $(x + 7)(x - 4)$
   d. $(x + 8)(x - 5)$

9. Factor the trinomial $x^2 + 2x - 80$ into a product of binomials.
   a. $(x + 20)(x - 18)$
   b. $(x + 10)(x - 8)$
   c. $(x + 16)(x - 5)$
   d. $(x + 8)(x - 10)$

10. Factor the trinomial $x^2 - 10x - 39$ into a product of binomials.
    a. $(x + 3)(x - 13)$
    b. $(x + 13)(x - 3)$
    c. $(x + 1)(x - 39)$
    d. $(x + 4)(x - 14)$

11. Factor the trinomial $x^2 - 15x - 16$ into a product of binomials.
    a. $(x + 16)(x - 1)$
    b. $(x + 1)(x - 16)$
    c. $(x + 2)(x - 17)$
    d. $(x + 2)(x - 8)$

12. Factor the trinomial $x^2 - x - 30$ into a product of binomials.
    a. $(x + 4)(x - 5)$
    b. $(x + 3)(x - 10)$
    c. $(x + 6)(x - 5)$
    d. $(x + 5)(x - 6)$

13. Last year you managed a square ice-skating rink with an area of x^2 square meters. This year you have expanded the rink into a rectangle with an area of $x^2 + 12x + 20$ square meters. If the rectangle is longer than it is wide, write its area in the factored form, Area = length × width.
    a. $(x + 20)(x + 1)$ square meters
    b. $(x + 8)(x + 4)$ square meters
    c. $(x + 5)(x + 4)$ square meters
    d. $(x + 10)(x + 2)$ square meters

14. You have a square piece of land with a total area of x^2 square feet. To prepare it for farming, you fence it in and dig a drainage ditch, reducing the farmable area to a rectangle with area $x^2 - 32x + 175$ square feet. If the rectangle is longer than it is wide, write its area in the factored form, Area = length × width.
    a. $(x - 1)(x - 175)$ square feet
    b. $(x - 5)(x - 35)$ square feet
    c. $(x - 7)(x - 25)$ square feet
    d. $(x - 8)(x - 24)$ square feet

15. You are replacing a square window in your kitchen with a rectangular window that is taller but with a narrower base. The original window has an area of x^2 square centimeters, and the new window has an area of $x^2 + x - 72$ square centimeters. Write the area of the new window in the factored form, Area = base × height.

   a. $(x-4)(x+18)$ square centimeters
   b. $(x-9)(x+8)$ square centimeters
   c. $(x-8)(x+9)$ square centimeters
   d. $(x-6)(x+12)$ square centimeters

6.10 Factoring the Difference of Two Squares

Practice Questions

1. Determine whether the binomial $-25 + m^2$ can be expressed as a difference of perfect squares of variables and integers. If so, factor it into a product of linear binomials.
 a. $(m-5)^2$
 b. $(m+25)(m-25)$
 c. $(m+5)(m-5)$
 d. The binomial is not the difference of perfect squares of variables and integers.

2. Determine whether the binomial $100 - x^2$ can be expressed as a difference of perfect squares of variables and integers. If so, factor it into a product of linear binomials.
 a. $(10-x)^2$
 b. $(10+x)(10-x)$
 c. $(25+x)(4-x)$
 d. The binomial is not the difference of perfect squares of variables and integers.

3. Determine whether the binomial $x^2 + 16$ can be expressed as a difference of perfect squares of variables and integers. If so, factor it into a product of linear binomials.
 a. $(x+4)^2$
 b. $(x+16)^2$
 c. $(x+4)(x-4)$
 d. The binomial is not the difference of perfect squares of variables and integers.

4. Determine whether the binomial $t^2 - 49$ can be expressed as a difference of perfect squares of variables and integers. If so, factor it into a product of linear binomials.
 a. $(t+49)(t-49)$
 b. $(t+7)(t-7)$
 c. $(t-7)^2$
 d. The binomial is not the difference of perfect squares of variables and integers.

5. Determine whether the binomial $x^2 - 1$ can be expressed as a difference of perfect squares of variables and integers. If so, factor it into a product of linear binomials.
 a. $(x-1)^2$
 b. $(x+1)^2$
 c. $(x+1)(x-1)$
 d. The binomial is not the difference of perfect squares of variables and integers.

6. Determine whether the binomial $x^2 - 6$ can be expressed as a difference of perfect squares of variables and integers. If so, factor it into a product of linear binomials.
   a. $(x + 2)(x - 3)$
   b. $(x + 6)(x - 6)$
   c. $(x - 3)^2$
   d. The binomial is not the difference of perfect squares of variables and integers.

7. Determine whether the binomial $25x^2 - 64$ can be expressed as a difference of perfect squares of variables and integers. If so, factor it into a product of linear binomials.
   a. $(5x + 32)(5x - 32)$
   b. $(5x - 8)^2$
   c. $(5x + 8)(5x - 8)$
   d. The binomial is not the difference of perfect squares of variables and integers.

8. Determine whether the binomial $x^2 - 36y^2$ can be expressed as a difference of perfect squares of variables and integers. If so, factor it into a product of linear binomials.
   a. $(x + 6y)(x - 6y)$
   b. $(x - 6y)^2$
   c. $(x + 4y)(x - 9y)$
   d. The binomial is not the difference of perfect squares of variables and integers.

9. Determine whether the binomial $4x^2 - 81$ can be expressed as a difference of perfect squares of variables and integers. If so, factor it into a product of linear binomials.
   a. $(2x - 9)^2$
   b. $(x + 3)(4x - 27)$
   c. $(2x + 9)(2x - 9)$
   d. The binomial is not the difference of perfect squares of variables and integers.

10. Determine whether the binomial $16p^2 - 25q^2$ can be expressed as a difference of perfect squares of variables and integers. If so, factor it into a product of linear binomials.
    a. $(4p - 5q)^2$
    b. $(4p + 5q)(4p - 5q)$
    c. $(16p + q)(p - 25q)$
    d. The binomial is not the difference of perfect squares of variables and integers.

11. Determine whether the binomial $9x^2 - 144$ can be expressed as a difference of perfect squares of variables and integers. If so, factor it into a product of linear binomials.
    a. $(3x - 12)^2$
    b. $(3x + 12)(3x - 12)$
    c. $(x + 9)(9x - 16)$
    d. The binomial is not the difference of perfect squares of variables and integers.

12. Determine whether the binomial $x^2 - 35y^2$ can be expressed as a difference of perfect squares of variables and integers. If so, factor it into a product of linear binomials.
    a. $(x + 5y)(x - 7y)$
    b. $(x - 5y)^2$
    c. $(x + 35y)(x - 35y)$
    d. The binomial is not the difference of perfect squares of variables and integers.

13. You have a collection of x^2 small square tiles all the same size, arranged in a large square. A rock falls on the square and breaks 9 of the tiles. You want to rearrange all the remaining tiles into a smaller square or rectangle. Write the number of undamaged tiles as a difference of perfect squares then factor it to get possible dimensions of the new square or rectangle.
    a. $(x + 9)(x - 9)$
    b. $(x - 3)(x - 3)$
    c. $(x + 3)(x - 3)$
    d. It is impossible to rearrange the remaining tiles into a square or rectangle.

14. You have a large, square recreation room, the floor of which is carpeted by $4x^2$ old square carpet tiles, each measuring 1 square foot. You decide to pull up these old tiles and to reuse them to replace the even older flooring in your rectangular den. When you finish, you have 100 of the old carpet tiles left over. Write the number of old carpet tiles used as a difference of perfect squares then factor it to get possible dimensions of the floor of the recreation room.
    a. $(2x + 4)(2x - 25)$
    b. $(4x + 100)(x - 1)$
    c. $(2x - 10)(2x - 10)$
    d. $(2x + 10)(2x - 10)$

15. A large baking sheet holds x^2 freshly baked chocolate chip cookies laid out in a square. After you eat 4 of them, you rearrange the remaining cookies in a smaller square or rectangle. Write the number of uneaten cookies as a difference of perfect squares then factor it to get possible dimensions of the new square or rectangle.
    a. $(x + 4)(x - 4)$
    b. $(x - 4)(x - 4)$
    c. $(x + 2)(x - 2)$
    d. $(x - 2)(x - 2)$

6.11 AC Method

PRACTICE QUESTIONS

1. Factor the trinomial $6x^2 + 5x + 1$ into a product of binomials.
    a. $(6x + 1)(x + 1)$
    b. $(3x + 1)(x + 2)$
    c. $(2x + 3)(x + 1)$
    d. $(2x + 1)(3x + 1)$

2. Factor the trinomial $5x^2 + 36x + 7$ into a product of binomials.
    a. $(x + 7)(5x + 1)$
    b. $(7x + 1)(5x + 1)$
    c. $(5x + 7)(x + 1)$
    d. $(7x + 1)(x + 5)$

3. Factor the trinomial $4x^2 + 16x + 15$ into a product of binomials.
    a. $(2x + 15)(2x + 1)$
    b. $(4x + 15)(x + 1)$
    c. $(4x + 3)(x + 5)$
    d. $(2x + 3)(2x + 5)$

4. Factor the trinomial $10x^2 - 17x + 3$ into a product of binomials.
    a. $(10x - 3)(x - 1)$
    b. $(10x - 1)(x - 3)$
    c. $(5x - 3)(2x - 1)$
    d. $(2x - 3)(5x - 1)$

5. Factor the trinomial $7x^2 - 18x + 8$ into a product of binomials.
    a. $(x - 2)(7x - 4)$
    b. $(x - 4)(7x - 2)$
    c. $(x - 1)(7x - 8)$
    d. $(x - 8)(7x - 1)$

6. Factor the trinomial $10x^2 - 29x + 10$ into a product of binomials.
    a. $(10x - 5)(x - 2)$
    b. $(5x - 5)(2x - 2)$
    c. $(2x - 5)(5x - 2)$
    d. $(5x - 1)(2x - 10)$

7. Factor the trinomial $14x^2 - 5x - 1$ into a product of binomials.
    a. $(2x - 1)(7x + 1)$
    b. $(x - 1)(14x + 1)$
    c. $(14x - 1)(x + 1)$
    d. $(7x - 1)(2x + 1)$

8. Factor the trinomial $5x^2 - 54x - 11$ into a product of binomials.
    a. $(5x - 11)(x + 1)$
    b. $(x - 11)(5x + 1)$
    c. $(5x - 1)(x + 11)$
    d. $(x - 1)(5x + 11)$

9. Factor the trinomial $4x^2 + 8x - 21$ into a product of binomials.
    a. $(x - 3)(4x + 7)$
    b. $(2x - 7)(2x + 3)$
    c. $(2x - 1)(2x + 21)$
    d. $(2x - 3)(2x + 7)$

10. Factor the trinomial $30x^2 + 7x - 2$ into a product of binomials.
    a. $(10x - 2)(3x + 1)$
    b. $(15x - 2)(2x + 1)$
    c. $(6x - 1)(5x + 2)$
    d. $(2x - 1)(15x + 2)$

11. Factor the trinomial $6x^2 - 11x - 10$ into a product of binomials.
    a. $(3x - 2)(2x + 5)$
    b. $(3x + 2)(2x - 5)$
    c. $(6x - 5)(x + 2)$
    d. $(2x - 1)(3x + 10)$

12. Factor the trinomial $9x^2 - 12x - 5$ into a product of binomials.
    a. $(3x + 1)(3x - 5)$
    b. $(x - 1)(9x + 5)$
    c. $(9x - 1)(x + 5)$
    d. $(3x - 1)(3x + 5)$

13. Last year you constructed a square corn maze with an area of x^2 square meters. This year you have expanded the maze into a rectangle that is longer than it is wide and whose area is $5x^2 + 11x + 6$ square meters. Factor this trinomial into a product of binomials to get a possible way of writing the area as length times width.

   a. $(5x + 2)(x + 3)$
   b. $(5x + 3)(x + 2)$
   c. $(5x + 6)(x + 1)$
   d. $(5x + 1)(x + 6)$

14. You are building a rectangular pen out of fencing panels x feet long ($x > 1$). Each side of the pen consists of one or more panels plus a small gap to accommodate a gate, and the length of the pen exceeds its width. The total area of the pen is $42x^2 + 19x + 2$ square feet. Factor this trinomial into a product of binomials to find a possible way of writing the area as length times width.

   a. $(7x + 1)(6x + 2)$
   b. $(7x + 2)(6x + 1)$
   c. $(14x + 1)(3x + 2)$
   d. $(14x + 2)(3x + 1)$

15. You are replacing a square window with area x^2 square inches ($x \geq 4$) by a larger rectangular window with area $6x^2 + 5x - 21$ square inches. The base of the new window is larger than its height. Factor this trinomial into a product of binomials to get a possible way of writing its area as base times height.

   a. $(3x + 3)(2x - 7)$
   b. $(3x + 21)(2x - 1)$
   c. $(3x + 7)(2x - 3)$
   d. $(6x + 1)(x - 7)$

6.12 Multistep Factoring

PRACTICE QUESTIONS

1. Factor the polynomial $-2x^5 - 24x^4 - 72x^3$ completely.
    a. $x^3(2x-6)^2$
    b. $-2x^3(x+6)^2$
    c. $-2x^3(x+4)(x+9)$
    d. $2x^5(x-4)(x-9)$

2. Factor the polynomial $4t^2 + 32t - 192$ completely.
    a. $4(t+4)(t-12)$
    b. $4(t+12)(t-4)$
    c. $4(t+8)(t-6)$
    d. $2(t+24)(2t-4)$

3. Factor the polynomial $20x^3 - 22x^2 + 6x$ completely.
    a. $2x(2x-1)(5x-3)$
    b. $2x(2x-3)(5x-1)$
    c. $2x(10x-3)(x-1)$
    d. $2x(10x-1)(x-3)$

4. Factor the polynomial $-21x^5 + 90x^4 - 24x^3$ completely.
    a. $-3x^3(x-2)(7x-4)$
    b. $-3x^3(x+2)(7x-4)$
    c. $-3x^3(x+4)(7x-2)$
    d. $-3x^3(x-4)(7x-2)$

5. Factor the polynomial $8x^6 - 96x^5 + 160x^4$ completely.
    a. $4x^4(2x-2)(x-20)$
    b. $8x^4(x-4)(x-5)$
    c. $8x^4(x-2)(x-10)$
    d. $4x^4(x-4)(2x-10)$

6. Factor the polynomial $40x^7 - 490x^5y^2$ completely.
    a. $40x^5y^2(x+7)(x-7)$
    b. $10x^5(2x-7y)^2$
    c. $10x^5(2x+7y)(2x-7y)$
    d. $10x^5(4x+y)(x-49y)$

7. Factor the polynomial $20x^{10} + 125x^9 - 105x^8$ completely.
    a. $5x^8(2x - 7)(2x + 3)$
    b. $5x^8(x + 7)(4x - 3)$
    c. $5x^8(x - 3)(4x + 7)$
    d. $5x^8(2x - 3)(2x + 7)$

8. Factor the polynomial $5x^3 + 45x^2 + 90x$ completely.
    a. $x(5x + 3)(x + 30)$
    b. $5x(x + 1)(x + 18)$
    c. $5x(x + 3)(x + 6)$
    d. $5x(x + 2)(x + 9)$

9. Factor the polynomial $90x^4 - 51x^3 - 6x^2$ completely.
    a. $3x^2(6x + 1)(5x - 2)$
    b. $3x^2(10x + 1)(3x - 2)$
    c. $3x^2(5x + 2)(6x - 1)$
    d. $6x^2(5x + 1)(3x - 1)$

10. Factor the polynomial $x^4 - 4x^3 - 45x^2$ completely.
    a. $x^2(x + 5)(x - 9)$
    b. $x^3(x + 5)(x - 9)$
    c. $x^2(x + 9)(x - 5)$
    d. $x^2(x + 15)(x - 3)$

11. Factor the polynomial $15x^4 + 36x^3 + 21x^2$ completely.
    a. $3x^2(x + 7)(5x + 1)$
    b. $3x^2(x + 1)(5x + 7)$
    c. $x^2(3x + 1)(5x + 21)$
    d. $x^2(x + 3)(15x + 7)$

12. Factor the polynomial $6x^2 - 486$ completely.
    a. $3(x + 18)(2x - 9)$
    b. $6(x - 9)^2$
    c. $6(x + 9)(x - 9)$
    d. $3(2x + 9)(x - 18)$

13. Your ice cream company used to sell ice cream in a cube-shaped package with volume x^3 cubic millimeters. To increase profits, you reduce the length and width of your package while leaving the height (and the price) the same. Your new package is a rectangular solid with a volume of $x^3 - 18x^2 + 72x$ cubic millimeters. Assume the length is greater than the width and you can find the dimensions of the new package by factoring the volume completely as a polynomial, rewrite the volume of the package in the form $V = $ length $\times$ width $\times$ height.
    a. $V = (x-4)(x-18)x$ cubic millimeters
    b. $V = (x-2)(x-36)x$ cubic millimeters
    c. $V = (x-6)(x-12)x$ cubic millimeters
    d. $V = (x-8)(x-9)x$ cubic millimeters

14. Your old storage tank for rainwater runoff consists of a cube-shaped hole in the ground with volume x^3 cubic meters. You expand it into a larger rectangular solid by increasing the length and width and doubling the depth, getting a new storage tank with volume $2x^3 + 58x^2 + 200x$ cubic meters. Assume the length is at least as large as the width and you can find the dimensions of the new tank by factoring the volume completely as a polynomial, rewrite the volume of the new tank in the form $V = $ length $\times$ width $\times$ depth.
    a. $V = (x+10)(x+10)(2x)$ cubic meters
    b. $V = (2x+25)(x+8)(x)$ cubic meters
    c. $V = (x+20)(x+5)(2x)$ cubic meters
    d. $V = (x+25)(x+4)(2x)$ cubic meters

15. A solar farm has a field entirely covered with solar cells. The field used to be square with an area of x^2 square feet, but recently the farm expanded the field into a larger rectangle. Each solar cell is the same size, and each square foot of the field holds the same number of cells. The total number of solar cells in the expanded field is $4x^2 + 96x + 512$. On this farm, as it happens, if you factor this polynomial completely, the factorization will express the total number of solar cells as (number of cells per square foot) $\times$ (length of the field in feet) $\times$ (width of the field in feet). Find this factorization.
    a. $4(x+16)(x+8)$
    b. $4(x+4)(x+32)$
    c. $4(x+2)(x+32)$
    d. $2(2x+8)(x+32)$

Chapter 7: Algebraic Fractions

7.1 Simplifying Algebraic Fractions

PRACTICE QUESTIONS

1. Simplify the following algebraic fraction.
$$\frac{6x^6}{30x^3}$$

   a. $\frac{3x^3}{15}$
   b. $\frac{x^2}{5}$
   c. $\frac{3x^2}{15}$
   d. $\frac{x^3}{5}$

2. Simplify the following algebraic fraction.
$$\frac{48x^4y^2}{12xy^5}$$

   a. $\frac{4x^3}{y^3}$
   b. $\frac{32}{5}$
   c. $\frac{4x^4}{y^3}$
   d. $\frac{4x^4}{y^5}$

3. Simplify the following algebraic fraction.
$$\frac{15x^3}{20x^{12}}$$

   a. $\frac{1}{5x^9}$
   b. $\frac{3}{4x^9}$
   c. $\frac{9}{40}$
   d. $\frac{3}{4x^4}$

4. Simplify the following algebraic fraction.
$$\frac{35x^2y^3}{14x^6y^6}$$

a. $\frac{21}{x^4y^3}$

b. $\frac{5}{12}$

c. $\frac{5}{2x^4y^3}$

d. $\frac{5}{2x^3y^2}$

5. Simplify the following algebraic fraction.
$$\frac{3x^3 + 12x^2}{5x^4 - 35x^3}$$

a. $-\frac{1}{2x}$

b. $\frac{3(x+4)}{5(x-7)}$

c. $-\frac{1}{2x^2}$

d. $\frac{3(x+4)}{5x(x-7)}$

6. Simplify the following algebraic fraction.
$$\frac{10x^4y + 40x^3y}{10x^2y^2 - 5xy^3}$$

a. $\frac{x^2(x+8)}{y(x-y)}$

b. $\frac{9x^2}{y-y^2}$

c. $\frac{6x^2}{y}$

d. $\frac{2x^2(x+4)}{y(2x-y)}$

7. Simplify the following algebraic fraction.
$$\frac{6a^5b - 18ab}{18a^2b^3 + 180a^2b^2}$$

a. $\frac{a^4-3}{3ab(b+10)}$

b. $\frac{a^3-1}{b(3b+10a)}$

c. $\frac{a^3-3}{3b(b+10)}$

d. $\frac{a^4-3}{12(b+10)}$

8. Simplify the following algebraic fraction.
$$\frac{x^2 + 3x}{3x + 9}$$

a. $\frac{x^2}{3}$

b. $\frac{x^2}{9}$

c. $\frac{x^2+3}{12}$

d. $\frac{x}{3}$

9. Simplify the following algebraic fraction.
$$\frac{2x^2 - 50}{6x^2 + 60x + 150}$$

a. $\frac{(x-5)^2}{3(x+5)^2}$

b. $\frac{-1}{3(10x+1)}$

c. $\frac{x-5}{3(x+5)}$

d. $\frac{2x}{3(x+5)}$

10. Simplify the following algebraic fraction.
$$\frac{16x^3 + 144x^2 + 320x}{20x^3 + 20x^2 - 240x}$$

a. $\frac{4(x+10)}{5(x-6)}$

b. $\frac{4(x+4)(x+5)}{5(x+3)(x-4)}$

c. $\frac{44}{7}$

d. $\frac{4(x+5)}{5(x-3)}$

11. Simplify the following algebraic fraction.
$$\frac{4x^3 - 20x^2}{30x - 6x^2}$$

a. $-\frac{2x}{3}$

b. $\frac{2x}{3}$

c. $\frac{(x^2-5)}{6}$

d. $\frac{2x(x-5)}{3(5-x)}$

12. Simplify the following algebraic fraction.
$$\frac{42x^2y^2 - 21x^3y}{14x^2y^3 - 28xy^4}$$

   a. $\frac{3x(2y-x)}{2y^2(x-2y)}$

   b. $-\frac{3x}{2y^2}$

   c. $\frac{3(1+x)(1-x)}{y(1+2y)(1-2y)}$

   d. $\frac{3x}{2y^2}$

13. A rectangular solid has a volume of $2x^3 + 20x^2 + 48x$, and its base has an area of $2x^2 + 8x$. Find the ratio of the solid's base area to its volume and simplify the ratio.

   a. $\frac{1}{x+6}$

   b. $\frac{2}{20x^2+x+6}$

   c. $\frac{x+4}{(x+2)(x+12)}$

   d. $\frac{x+8}{(x+3)(x+16)}$

14. A circle with radius r has a circumference of $2\pi r$ and an area of πr^2. Find the ratio of its circumference to its area and simplify the ratio as much as possible.

   a. $\frac{2\pi}{r}$

   b. $\frac{2}{r}$

   c. $\frac{2}{\pi r}$

   d. $\frac{1}{2}$

15. A pyramid has a volume of $2x^4 + 12x^3 + 16x^2$, and its base has an area of $6x^3 + 24x^2$. Find the ratio of the pyramid's base area to its volume and simplify the ratio.

   a. $\frac{3(x+24)}{(x+2)(x+4)}$

   b. $\frac{5}{x(17x+1)}$

   c. $\frac{3}{x+2}$

   d. $\frac{3(x+4)}{(x+1)(x+8)}$

7.2 Solving Proportions Containing Algebraic Fractions

PRACTICE QUESTIONS

1. Solve the following proportion.
$$\frac{6}{x+6} = \frac{2}{3}$$

   a. $x = -3$
   b. $x = 15$
   c. $x = 3$
   d. $x = -\frac{9}{2}$

2. Solve the following proportion.
$$\frac{1}{2} = \frac{5}{7-5x}$$

   a. $x = 5$
   b. $x = -\frac{3}{5}$
   c. $x = 1$
   d. $x = \frac{3}{5}$

3. Solve the following proportion.
$$\frac{7}{3x-5} = \frac{4}{x+5}$$

   a. $x = \frac{1}{11}$
   b. $x = 2$
   c. $x = 3$
   d. $x = 11$

4. Solve the following proportion.
$$\frac{3}{x+2} = \frac{5}{x-10}$$

   a. $x = -20$
   b. $x = 20$
   c. $x = -5$
   d. $x = -10$

5. Solve the following proportion.

$$\frac{6-9x}{3-4x} = \frac{3}{2}$$

a. $x = -\frac{1}{2}$
b. $x = \frac{1}{2}$
c. $x = -\frac{1}{10}$
d. $x = 2$

6. Solve the following proportion.

$$\frac{2x+3}{2} = \frac{x+3}{3}$$

a. $x = -\frac{3}{4}$
b. $x = \frac{3}{4}$
c. $x = -\frac{15}{4}$
d. $x = -\frac{4}{3}$

7. Solve the following proportion.

$$\frac{1}{x+2} = \frac{x+3}{x^2+16}$$

a. $x = -3$
b. $x = 2$
c. $x = \frac{22}{5}$
d. $x = 5$

8. Solve the following proportion.

$$\frac{x+2}{x+6} = \frac{x-6}{x-4}$$

a. $x = \frac{1}{14}$
b. $x = -14$
c. $x = 22$
d. $x = 14$

9. Solve the following proportion.

$$\frac{2x-4}{x-3} = \frac{x+7}{x+5}$$

a. $x = -1$
b. $x = 1$
c. $x = 4$
d. $x = -2$

10. Solve the following proportion.
$$\frac{x+6}{2x+6} = \frac{x-2}{x+10}$$

   a. $x = -8$ or $x = 9$
   b. $x = -3$ or $x = 24$
   c. $x = -6$ or $x = 12$
   d. $x = -4$ or $x = 18$

11. Solve the following proportion.
$$\frac{x+3}{x+8} = \frac{x+5}{2x+4}$$

   a. $x = -14$ or $x = 2$
   b. $x = -4$ or $x = 7$
   c. $x = -2$ or $x = 14$
   d. $x = -7$ or $x = 4$

12. Solve the following proportion.
$$\frac{-x-16}{x+4} = \frac{x-8}{x+2}$$

   a. $x = -8$ or $x = 16$
   b. $x = 2$ or $x = 4$
   c. $x = 0$ or $x = 7$
   d. $x = -7$ or $x = 0$

13. A pitcher of lemonade is diluted with 2 cups of water to get 5 cups of lemonade. You want to dilute the mixture with additional water until water makes up $\frac{7}{13}$ of the lemonade. Let x be the number of cups of water you need to add. Set up and solve a proportion of algebraic fractions to find x.

   a. $3\frac{1}{2}$ cups
   b. $\frac{1}{2}$ cup
   c. $1\frac{1}{2}$ cups
   d. $2\frac{1}{2}$ cups

14. A school assembly has x boys and x girls in the audience. After 4 boys and 8 girls leave to go on a field trip, the ratio of boys to girls becomes $\frac{16}{15}$. Set up and solve a proportion of algebraic fractions to find x.

   a. $x = 17$
   b. $x = 56$
   c. $x = 68$
   d. $x = 46$

15. You are trying to mix up an oil-and-vinegar salad dressing. You begin by mixing 17 cups of olive oil with 4 cups of vinegar, but the dressing is not tart enough. To fix this, you let the dressing separate, pour off x cups of olive oil from the top, and then add x cups of vinegar to replace it. In your final mixture, which tastes perfect, the ratio of oil to vinegar is $\frac{5}{2}$. Set up and solve a proportion of algebraic fractions to find x.

   a. $3\frac{1}{2}$ cups
   b. 2 cups
   c. 3 cups
   d. $2\frac{1}{2}$ cups

7.3 Multiplying Algebraic Fractions

PRACTICE QUESTIONS

1. Calculate the following product. Simplify your answer.
$$\frac{x+5}{x+10} \cdot \frac{x+5}{x-10}$$

   a. $\frac{x^2+10x+25}{x^2-100}$
   b. $\frac{x^2+10x+25}{x^2-20x-100}$
   c. $\frac{x^2+25}{x^2-100}$
   d. $\frac{x^2+25}{x^2-20x-100}$

2. Calculate the following product. Simplify your answer.
$$\frac{3}{5x} \cdot \frac{10x^2+15x}{6x+3}$$

   a. $\frac{x+3}{x+1}$
   b. $\frac{17x}{2x+3}$
   c. $\frac{2x+3}{2x+1}$
   d. 2

3. Calculate the following product. Simplify your answer.
$$\frac{3x}{4x-8} \cdot \frac{14x-28}{12x+15}$$

   a. $\frac{7x^2-14x}{8x^2-6x-20}$
   b. $\frac{42x^2-84x}{48x^2-36x-120}$
   c. $\frac{7x-8}{-8x-10}$
   d. $\frac{7x}{8x+10}$

4. Calculate the following product. Simplify your answer.
$$\frac{4x+1}{30} \cdot \frac{20x-30}{2x+11}$$

   a. $\frac{80x^2+20x}{2x+11}$
   b. $\frac{40x^2-130x-30}{30x+330}$
   c. $\frac{8x^2-10x-3}{6x+33}$
   d. $\frac{4x^2-4x-3}{6x+33}$

5. Calculate the following product. Simplify your answer.
$$\frac{20-5x}{10x+11} \cdot \frac{10}{7x-28}$$

a. $\frac{190}{91x-234}$
b. $\frac{-50x+20}{7x^2+49x-308}$
c. $\frac{-50}{70x+77}$
d. $\frac{-5x+20}{7x^2+49x-308}$

6. Calculate the following product. Simplify your answer.
$$\frac{9x+6}{8x-40} \cdot \frac{4x-20}{15x+10}$$

a. $\frac{9x+6}{30x+20}$
b. $\frac{3}{10}$
c. $\frac{9x^2+39x-30}{30x^2-130x-100}$
d. $\frac{18x^2-3x-3}{60x^2-146x-10}$

7. Calculate the following product. Simplify your answer.
$$\frac{7x+3}{-2x^2+16x-30} \cdot \frac{15-3x}{4}$$

a. $\frac{21x+9}{8x-24}$
b. $\frac{9+21x}{24-8x}$
c. $\frac{-21x^2+96x+45}{-8x^2+64x-120}$
d. $\frac{21x^2+3x+3}{8x^2+2x-8}$

8. Calculate the following product. Simplify your answer.
$$\frac{2x^2+20x+18}{x^2-12x+11} \cdot \frac{x-11}{4x+36}$$

a. $\frac{x+1}{2x-2}$
b. $\frac{x^2-10x-11}{2x^2-24x+22}$
c. $\frac{x^2+10x+9}{2x^2+16x-18}$
d. $\frac{2x+2}{4x-4}$

9. Calculate the following product. Simplify your answer.
$$\frac{9x}{5x^2 + 10x - 40} \cdot \frac{-2x^2 - 4x + 16}{6x^2 - 39x}$$

a. $\frac{-6}{10x-65}$
b. $\frac{3x}{2x-13}$
c. $\frac{-6x-24}{10x^2-25x-260}$
d. $\frac{12-6x}{10x^2-85x+130}$

10. Calculate the following product. Simplify your answer.
$$\frac{12x^2 + 36x}{x^2 + 7x + 10} \cdot \frac{x^2 + 9x + 20}{8x^2 + 24x}$$

a. $\frac{3x^2+21x+36}{2x^2+10x+12}$
b. $\frac{3x+12}{2x+4}$
c. $\frac{3x+3}{2x+1}$
d. $\frac{3x^2+27x+60}{2x^2+19x+20}$

11. Calculate the following product. Simplify your answer.
$$\frac{x^2 + 6x - 7}{x^2 - x - 20} \cdot \frac{x^2 - 2x - 15}{x^2 + 5x - 14}$$

a. $\frac{x^2+2x-3}{x^2+2x-8}$
b. $\frac{x^2+2x-3}{x^2-6x+8}$
c. $\frac{x^2-4x+3}{x^2+2x-8}$
d. $\frac{x^2-4x+3}{x^2-6x+8}$

12. Calculate the following product. Simplify your answer.
$$\frac{x^3 - 7x^2 - 8x}{x^2 - 8x + 12} \cdot \frac{x^2 - 3x + 2}{x^3 - 2x^2 - 3x}$$

a. $\frac{x^3-9x^2+8x}{x^3-9x^2+18x}$
b. $\frac{x^2-x+4}{x^2-x+9}$
c. $\frac{x^2-9x+8}{x^2-9x+18}$
d. $\frac{x^2-5x+4}{x^2-6x+9}$

13. On a particular day, a clothing manufacturer produces a large number of socks. The socks are inspected for quality, and the fraction with noticeable flaws is $\frac{x-4}{3x^2+6x}$. Of these socks with flaws, the fraction good enough to be sold at a discount as "factory seconds" is $\frac{3x}{x^2-16}$, with the rest being discarded. What fraction of the day's production will be sold as factory seconds? Write your answer as a reduced algebraic fraction.

    a. $\frac{1}{x^2+6x+8}$
    b. $\frac{12}{9x^2-144x}$
    c. $\frac{3x^2-12x}{4x^2+6x-16}$
    d. $\frac{6x}{x^2-16}$

14. One day on the stock market, the fraction of stocks that increased in price was $\frac{9x+36}{16x+32}$. Of these, the fraction that increased in price again the next day was $\frac{8}{5x+20}$. What fraction of stocks increased on both days? Write your answer as a reduced algebraic fraction.

    a. $\frac{72x+288}{80x^2+480x+640}$
    b. $\frac{6}{5x+20}$
    c. $\frac{9}{10x+20}$
    d. $\frac{72x+288}{80x^2+640}$

15. In a clinical trial of vaccines, the fraction of volunteers receiving Vaccine A was $\frac{5x}{4x^2+20x+16}$. Of these, the fraction developing therapeutic levels of antibodies after three weeks was $\frac{2x+8}{3x^2+6x}$. What fraction of the original group of volunteers both received Vaccine A and developed therapeutic levels of antibodies after three weeks? Write your answer as a reduced algebraic fraction.

    a. $\frac{5}{24x^2+12}$
    b. $\frac{5}{6x^2+12}$
    c. $\frac{5}{6x^2+18x+12}$
    d. $\frac{5}{27x^2+18x}$

7.4 Dividing Algebraic Fractions

PRACTICE QUESTIONS

1. Calculate the following quotient. Simplify your answer.
$$\frac{x+4}{x+8} \div \frac{x-8}{x+4}$$

   a. $\frac{x^2+16}{x^2-64}$
   b. $\frac{x^2+8x+1}{x^2-4}$
   c. $\frac{x^2+8x+16}{x^2-16x-64}$
   d. $\frac{x^2+8x+16}{x^2-64}$

2. Calculate the following quotient. Simplify your answer.
$$\frac{2}{7x} \div \frac{6x+8}{21x^2+35x}$$

   a. $\frac{3x+5}{3x+4}$
   b. $\frac{x^2+5x}{x^2+4x}$
   c. $\frac{42x^2+70x}{42x^2+56x}$
   d. $\frac{x+5}{x+4}$

3. Calculate the following quotient. Simplify your answer.
$$\frac{5x}{6x-42} \div \frac{10x+15}{21x-147}$$

   a. $\frac{7x}{4x+6}$
   b. $\frac{7}{10}$
   c. $\frac{-42x}{70x-490}$
   d. $\frac{21x^2-147x}{12x^2-66x-126}$

4. Calculate the following quotient. Simplify your answer.
$$\frac{6x+1}{15} \div \frac{3x+10}{15x-10}$$

   a. $\frac{10x^2+3x-1}{5x+5}$
   b. $\frac{2x^2-9x-2}{x+30}$
   c. $\frac{18x^2-9x-2}{9x+30}$
   d. $\frac{18x^2-x-2}{x+30}$

5. Calculate the following quotient. Simplify your answer.
$$\frac{18-6x}{7x+2} \div \frac{5x-15}{7}$$

   a. $\frac{-42}{35x+10}$
   b. $\frac{126-42x}{x^2-19x-6}$
   c. $\frac{-6}{5x+10}$
   d. $\frac{126+7x}{7x^2-19x+1}$

6. Calculate the following quotient. Simplify your answer.
$$\frac{4x+6}{9x-36} \div \frac{14x+21}{3x-12}$$

   a. $\frac{2x+2}{7x+7}$
   b. $\frac{2}{21}$
   c. $\frac{12x+18}{42x+63}$
   d. $\frac{2x-8}{21x-84}$

7. Calculate the following quotient. Simplify your answer.
$$\frac{5x+4}{-3x^2+18x-24} \div \frac{5}{8-2x}$$

   a. $\frac{8-2x}{6-3x}$
   b. $\frac{2x+4}{3x-15}$
   c. $\frac{10x+8}{15x-30}$
   d. $\frac{2x+8}{3x-6}$

8. Calculate the following quotient. Simplify your answer.
$$\frac{3x^2+15x+18}{x^2-9x+14} \div \frac{6x+18}{x-7}$$

   a. $\frac{x^2+5x+6}{2x^2+2x-12}$
   b. $\frac{x^2-5x-14}{2x^2-18x+28}$
   c. $\frac{x+2}{2x-4}$
   d. $\frac{x+1}{2x-2}$

9. Calculate the following quotient. Simplify your answer.
$$\frac{8x}{3x^2 + 6x - 45} \div \frac{12x^2 - 44x}{-5x^2 - 10x + 75}$$

a. $\frac{-10x}{9x^2 - 33x}$
b. $\frac{30 - 10x}{9x^2 - 60x + 99}$
c. $\frac{-10x - 50}{9x^2 + 12x - 165}$
d. $\frac{-10}{9x - 33}$

10. Calculate the following quotient. Simplify your answer.
$$\frac{9x^2 + 9x}{x^2 + 6x + 8} \div \frac{6x^2 + 6x}{x^2 + 5x + 6}$$

a. $\frac{3x^2 + 24x + 45}{2x^2 + 18x + 40}$
b. $\frac{3x + 9}{2x + 8}$
c. $\frac{3x^2 + 18x + 27}{2x^2 + 14x + 24}$
d. $\frac{9x + 27}{6x + 24}$

11. Calculate the following quotient. Simplify your answer.
$$\frac{x^2 + 4x - 12}{x^2 - x - 12} \div \frac{x^2 + 5x - 6}{x^2 - 2x - 8}$$

a. $\frac{x^2 + 8x + 12}{x^2 + 6x + 9}$
b. $\frac{x^2 - 4x + 4}{x^2 + 2x - 3}$
c. $\frac{x^2 - 4}{x^2 + 2x - 3}$
d. $\frac{-3}{2x - 2}$

12. Calculate the following quotient. Simplify your answer.
$$\frac{x^3 - 6x^2 - 7x}{x^2 - 9x + 20} \div \frac{x^3 - 2x^2 - 3x}{x^2 - 7x + 10}$$

a. $\frac{x^2 - 9x + 7}{x^2 - 7x + 6}$
b. $\frac{x^2 - 12x + 35}{x^2 - 13x + 30}$
c. $\frac{x^3 - 9x^2 + 14x}{x^3 - 7x^2 + 12x}$
d. $\frac{x^2 - 9x + 14}{x^2 - 7x + 12}$

13. A red rectangle has an area of $x^2 + 7x + 10$ and a length of $x + 1$. A blue rectangle has an area of $x^2 + 9x + 20$ and a length of $x + 3$. Write an algebraic expression for the width of each rectangle. Then find an expression for the ratio of the width of the red rectangle to the width of the blue rectangle. Simplify your answer.

   a. $\dfrac{x^2+5x+6}{x^2+5x+4}$
   b. $\dfrac{8x+10}{5x+10}$
   c. $\dfrac{x^2+x+3}{x^2+x+2}$
   d. $\dfrac{x+3}{x+2}$

14. You sell $x + 4$ widgets and make a total gross income of $100. You sell $x + 8$ gadgets and make a total gross income of $200. Write an algebraic expression for the price of a widget and one for the price of a gadget. Then, find an expression for the ratio of the price of a widget to the price of a gadget. Simplify your answer.

   a. $\dfrac{x+8}{2x+8}$
   b. $\dfrac{20{,}000}{x^2+12x+32}$
   c. $\dfrac{2x+8}{x+8}$
   d. $\dfrac{x+1}{2x+1}$

15. A Christmas present is in a box, a rectangular solid, with a volume of $10x^2 + 50x$. Its base has an area of $x^2 + 11x + 10$. A birthday present is in a box, also a rectangular solid, with a volume of $15x^2 + 75x$. Its base has an area of $x^2 + 13x + 30$. Find an algebraic expression for the height of each box. Then, find an expression for the ratio of the height of the Christmas box to the height of the birthday box. Simplify your answer. (Hint: Volume of rectangular solid = area of base × height)

   a. $\dfrac{2x+2}{3x+1}$
   b. $\dfrac{2}{3}$
   c. $\dfrac{2x+6}{3x+3}$
   d. $\dfrac{10x+30}{15x+15}$

7.5 Adding and Subtracting Algebraic Fractions

Practice Questions

1. Evaluate the following sum. Simplify your answer.
$$\frac{x^2 + 3x - 8}{5x} + \frac{7x - 12}{5x}$$

   a. $\frac{x^2 + 10x - 20}{10x}$
   b. $\frac{x^2 + 10x - 20}{5x}$
   c. $\frac{x^2 + 10x - 4}{x}$
   d. $x^2 - 18$

2. Evaluate the following difference. Simplify your answer.
$$\frac{3x^2 - 7x + 12}{2x^2} - \frac{x^2 + 3x + 12}{2x^2}$$

   a. $\frac{x^2 - 2x + 12}{x^2}$
   b. $\frac{x - 5}{x}$
   c. $1 - 10x$
   d. $\frac{2x^2 - 10x}{2x^2}$

3. Evaluate the following sum. Simplify your answer.
$$\frac{3x^2 + 30x - 19}{4x^2 + 12x} + \frac{5x^2 - 2x + 19}{4x^2 + 12x}$$

   a. $\frac{8x^2 + 28x}{4x^2 + 12x}$
   b. $\frac{2x + 7}{2x + 6}$
   c. $\frac{2x + 7}{x + 3}$
   d. $\frac{2x^2 + 7x}{x^2 + 3x}$

4. Evaluate the following difference. Simplify your answer.
$$\frac{4x^2 - 3x - 17}{6x^2 - 24x} - \frac{x^2 - 9x + 7}{6x^2 - 24x}$$

   a. $\frac{x^2 + 2x - 8}{2x^2 - 8x}$
   b. $\frac{x^2 + x - 1}{x^2 - x}$
   c. $\frac{3x^2 + x - 1}{x^2 - x}$
   d. $\frac{x - 2}{2x}$

5. Evaluate the following sum. Simplify your answer.
$$\frac{8x^2 - 30x - 12}{x^3 - 4x^2} + \frac{-2x^2 + 6x + 12}{x^3 - 4x^2}$$

a. $\frac{6x-6}{x^2-x}$
b. $\frac{6x-24}{x^2-4x}$
c. $\frac{6x^2-24x}{x^3-4x^2}$
d. $\frac{6}{x}$

6. Evaluate the following difference. Simplify your answer.
$$\frac{5x^2 + 10x + 20}{x^2 + 3x - 4} - \frac{3x^2 - 8x - 20}{x^2 + 3x - 4}$$

a. $\frac{2x^2+2x}{x^2+3x-4}$
b. $\frac{2x}{x-4}$
c. $\frac{2x+10}{x-1}$
d. $\frac{2x^2+6x+10}{x^2+x-4}$

7. Evaluate the following sum. Simplify your answer.
$$\frac{x+11}{5} + \frac{14}{x}$$

a. $\frac{x^2+11x+14}{x}$
b. $\frac{x+25}{x+5}$
c. $\frac{x^2+11x+70}{5x}$
d. $\frac{x+25}{5x}$

8. Evaluate the following difference. Simplify your answer.
$$\frac{x}{3} - \frac{5x^2 - 4}{3x}$$

a. $\frac{-4x^2+4}{3x}$
b. $\frac{-4x+4}{3}$
c. $\frac{-4x^2-4}{3x}$
d. $\frac{x-5x^2+4}{3-3x}$

9. Evaluate the following sum. Simplify your answer.
$$\frac{5x-2}{4x} + \frac{7x+3}{6x}$$

a. $\frac{29x^2}{12x^2}$
b. $\frac{29x-1}{6x}$
c. $\frac{12x+1}{10x}$
d. $\frac{29}{12}$

10. Evaluate the following difference. Simplify your answer.
$$\frac{10x^2+5x-9}{8x^2} - \frac{10x-4}{8x}$$

a. $\frac{9x-9}{4x^2}$
b. $\frac{x-9}{8x^2}$
c. $\frac{10x^2-5x-5}{8x^2-8x}$
d. $\frac{9x-9}{8x^2}$

11. Evaluate the following sum. Simplify your answer.
$$\frac{7x-9}{5x^2} + \frac{2x+4}{3x^3}$$

a. $\frac{9x-5}{5x^2+3x^3}$
b. $\frac{21x^2-17x+20}{15x^3}$
c. $\frac{21x^4-17x^3+20x^2}{15x^5}$
d. $\frac{21x^2-17x+40}{30x^3}$

12. Evaluate the following difference. Simplify your answer.
$$\frac{x^2-9x+3}{6x^4} - \frac{x^2-12x+4}{8x^4}$$

a. $\frac{x^2-3x+1}{x^2}$
b. $\frac{1-3x}{2x^4}$
c. $\frac{1}{24x^2}$
d. $\frac{x^2-72x+24}{24x^2}$

13. A rectangle has a length of $\frac{x}{5}$ and a width of $\frac{5}{x}$. Find the perimeter of the rectangle. Simplify your answer.

   a. $\frac{2x+10}{5+x}$
   b. $\frac{2x+10}{x}$
   c. $\frac{x^2+25}{5x}$
   d. $\frac{2x^2+50}{5x}$

14. One week your company manufactures a large batch of widgets at an average cost of $\frac{x^2+4x-21}{x+10}$ dollars per widget and sells them at an average price of $\frac{x^2+11x+24}{x+10}$ dollars per widget. Find an expression for your average dollars of profit per widget. Simplify your answer.
Hint: When the number of widgets produced equals the number of widgets sold, as it does here, the average profit equals the average price minus the average cost.

   a. $\frac{7x+45}{x+10}$
   b. $\frac{x^2+15x+3}{x+10}$
   c. 8
   d. $\frac{7x+9}{x+2}$

15. Two boxes (rectangular solids) both have bases with an area of $4x^2$. The taller box has a volume of $9x^2 + 11x - 10$, and the shorter box has a volume of $7x^2 + 7x - 10$. Find the difference in height between the taller box and the shorter one. Simplify your answer.
Hint: Volume of rectangular solid = area of base × height

   a. $\frac{x+1}{x}$
   b. $2 + x$
   c. $8x^4 - 16x^3$
   d. $\frac{x+2}{2x}$

7.6 Adding and Subtracting Algebraic Fractions with Binomial Denominators

Practice Questions

1. Simplify: $\frac{5}{x-1} + \frac{2}{x-1}$

   a. $\frac{3}{x-1}$
   b. $\frac{7}{x-1}$
   c. $\frac{7}{2x-2}$
   d. $\frac{7}{(x-1)^2}$

2. Simplify: $\frac{3}{y+1} + \frac{5}{y-1}$

   a. $\frac{8y+2}{2y}$
   b. $\frac{8}{2y}$
   c. $\frac{8}{(y+1)(y-1)}$
   d. $\frac{8y+2}{(y+1)(y-1)}$

3. Simplify: $\frac{6}{x+4} - \frac{x}{x+4}$

   a. $\frac{6-x}{(x+4)-(x+4)}$
   b. $\frac{6-x}{x+4}$
   c. $\frac{x-6}{x+4}$
   d. $\frac{6-x}{(x+4)^2}$

4. Simplify: $\frac{x}{x^2-9} - \frac{1}{x+3}$

   a. $\frac{3}{(x+3)(x-3)}$
   b. $\frac{x-1}{x^2-9}$
   c. $\frac{3}{x^2-x-12}$
   d. $\frac{-1}{-x^2+12}$

5. Simplify: $\frac{x}{x-2} + \frac{3x}{x-2}$

   a. $\frac{4x}{2x-4}$
   b. $\frac{4x}{x^2-4}$
   c. $\frac{4x}{x-2}$
   d. $\frac{x+3x}{(x-2)^2}$

6. Simplify: $\frac{4}{x+2} + \frac{8}{x-3}$

   a. $\frac{12}{2x-1}$
   b. $\frac{12x+4}{2x-1}$
   c. $\frac{12x+4}{(x+2)(x-3)}$
   d. $\frac{12}{(x+2)(x-3)}$

7. Simplify: $\frac{x-7}{6x+3} - \frac{x}{6x+3}$

   a. $\frac{-7}{1}$
   b. $\frac{-7}{(6x+3)^2}$
   c. $\frac{2x-7}{6x+3}$
   d. $\frac{-7}{6x+3}$

8. Simplify: $\frac{6}{x+4} - \frac{2}{x+1}$

   a. $\frac{4}{(x+4)(x+1)}$
   b. $\frac{4x-2}{(x+4)(x+1)}$
   c. $\frac{4}{3}$
   d. $\frac{4x-2}{3}$

9. Simplify: $\frac{2x+6}{x^2-5} + \frac{x+2}{x^2-5}$

   a. $\frac{3x+8}{x^2-5}$
   b. $\frac{3x+8}{2x^2-10}$
   c. $\frac{3x+8}{2(x^2-5)}$
   d. $\frac{8}{2x-10}$

10. Simplify: $\frac{6}{k^2-4} + \frac{k}{k+2}$

    a. $\frac{k^2-2k+6}{k^2-4}$
    b. $\frac{6}{k^2-4}$
    c. $\frac{2+k}{6}$
    d. $\frac{6+k}{k-2}$

11. Simplify: $\frac{6x-8}{4x^3+2} - \frac{7x+3}{4x^3+2}$

    a. $\frac{-x-11}{4x^3+2}$
    b. $\frac{-x-5}{4x^3+2}$
    c. $\frac{-x-11}{(4x^3+2)-(4x^3-2)}$
    d. $\frac{x+11}{4x^3+2}$

12. Simplify: $\frac{y}{y+1} - \frac{y}{y+3}$

    a. $\frac{2y}{2y+4}$
    b. $-y$
    c. $\frac{2y}{y^2+4y+3}$
    d. $\frac{2y}{y^2+4}$

13. Adam and Joey are filling a garden bed with soil. Adam shoveled in $\frac{3m}{m+4}$ ft^3, and Joey shoveled in $\frac{4m}{m+5}$ ft^3. How much soil did Adam and Joey shovel into the garden bed altogether?

    a. $\frac{-1m}{(m+4)(m+5)}$ ft^3
    b. $\frac{7m^2+31m}{(m+4)(m+5)}$ ft^3
    c. $\frac{1m}{(m+4)(m+5)}$ ft^3
    d. $\frac{7m^2+31m}{2m+9}$ ft^3

14. Monica is reupholstering a bench. She has $\frac{a^2}{a+8}$ yards of fabric. She uses $\frac{4}{a+8}$ yards of her fabric to cover the bench. How many yards of fabric is she left with?

   a. $\frac{a^2-4}{2a+16}$ yd
   b. $\frac{a^2+4}{2a+16}$ yd
   c. $\frac{a^2+4}{a+8}$ yd
   d. $\frac{a^2-4}{a+8}$ yd

15. Martin's bookstore received a shipment of two boxes. The first box weighed $\frac{5}{x+1}$ lb, and the second box weighed $\frac{2}{x^2-1}$ lb. How much did the shipment weigh in all?

   a. $\frac{3}{(x+1)(x-1)}$ lb
   b. $\frac{7}{(x+1)(x-1)}$ lb
   c. $\frac{-3x+5}{(x+1)(x-1)}$ lb
   d. $\frac{5x-3}{(x+1)(x-1)}$ lb

7.7 Solving Equations Involving Algebraic Fractions

PRACTICE QUESTIONS

1. Solve for x.

$$\frac{x+1}{3} + \frac{x-4}{2} = 10$$

   a. $x = 8$
   b. $x = 10$
   c. $x = 12$
   d. $x = 14$

2. Solve for x.

$$\frac{x}{2} = \frac{x}{4} + 2$$

   a. $x = 2$
   b. $x = 8$
   c. $x = 4$
   d. $x = 6$

3. Solve for x.

$$\frac{2x}{5} - \frac{x}{4} = 3$$

   a. $x = 20$
   b. $x = 30$
   c. $x = 3$
   d. $x = 2$

4. Solve for x.

$$\frac{x+1}{2} - 6 = \frac{x-3}{6}$$

   a. $x = 12$
   b. $x = 9$
   c. $x = 15$
   d. $x = 6$

5. Solve for x.

$$\frac{x-1}{3} + \frac{x+1}{2} = 1$$

   a. $x = 5$
   b. $x = -1$
   c. $x = 25$
   d. $x = 1$

6. Solve for x.

$$\frac{2x-2}{6} = 9 + \frac{2x-4}{8}$$

   a. $x = 1$
   b. $x = 2.5$
   c. $x = 106$
   d. $x = 105$

7. Solve for x.

$$\frac{x+8}{5} - \frac{x-6}{3} = 4$$

   a. $x = -57$
   b. $x = 57$
   c. $x = -3$
   d. $x = 3$

8. Solve for x.

$$\frac{x+2}{2} = 4 - \frac{x+4}{3}$$

   a. $x = 2$
   b. $x = 10$
   c. $x = 38$
   d. $x = \frac{17}{5}$

9. Solve for x.

$$\frac{x-1}{2} + \frac{2x+3}{4} = 1$$

   a. $x = 2$
   b. $x = \frac{3}{4}$
   c. $x = \frac{1}{2}$
   d. $x = 4$

10. Solve for x.

$$\frac{4x-2}{6} = 9 + \frac{4x-4}{8}$$

   a. $x = 208$
   b. $x = 59$
   c. $x = 49$
   d. $x = 53$

11. Solve for x.

$$\frac{x}{6} - \frac{2x}{3} = 3$$

   a. $x = 6$
   b. $x = -6$
   c. $x = \frac{18}{5}$
   d. $x = -\frac{18}{5}$

12. Solve for x.

$$\frac{2x-1}{4} = 2 - \frac{x-1}{5}$$

   a. $x = \frac{49}{6}$
   b. $x = \frac{7}{2}$
   c. $x = 3$
   d. $x = -3$

13. Sarah and Lucan are putting marbles into a bag. Sarah put $\frac{x+2}{2}$ marbles in the bag, and Lucan put $\frac{x+4}{8}$ marbles in the bag. Altogether they had 9 marbles. Solve for x to determine how many marbles Sarah and Lucan each put into the bag.
   a. Sarah put 5 marbles into the bag, and Lucan put 4 marbles into the bag.
   b. Sarah put 8 marbles into the bag, and Lucan put 1 marble into the bag.
   c. Sarah put 7 marbles into the bag, and Lucan put 2 marbles into the bag.
   d. Sarah put 3 marbles into the bag, and Lucan put 6 marbles into the bag.

14. Alex had $\frac{6x-1}{4}$ pencils in his backpack. After lending $\frac{5-2x}{2}$ pencils to Jasmine, Alex had 1 pencil remaining. How many pencils did Alex start with? How many did he lend to Jasmine?
   a. Alex started with 10 pencils and lent 9 pencils to Jasmine.
   b. Alex started with 3 pencils and lent 2 pencils to Jasmine.
   c. Alex started with 5 pencils and lent 4 pencils to Jasmine.
   d. Alex started with 2 pencils and lent 1 pencil to Jasmine.

15. There are $\frac{2x}{5}$ boys in Ms. Smith's class and $\frac{3x}{10}$ boys in Ms. Cole's class. Both classes combined have a total of 14 boys. Solve for x to identify the number of boys in each class.
   a. There are 5 boys in Ms. Smith's class and 9 boys in Ms. Cole's class.
   b. There are 10 boys in Ms. Smith's class and 4 boys in Ms. Cole's class.
   c. There are 8 boys in Ms. Smith's class and 6 boys in Ms. Cole's class.
   d. There is 1 boy in Ms. Smith's class and 13 boys in Ms. Cole's class.

Chapter 8: Quadratic Functions

8.1 Domain and Range of Quadratic Functions

PRACTICE QUESTIONS

1. Find the domain of the quadratic function f whose graph appears here.

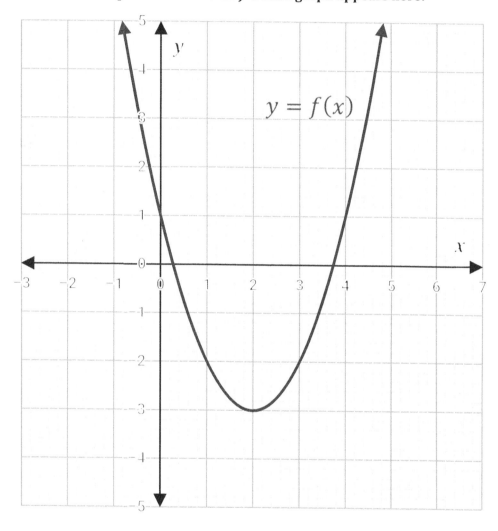

a. $-\infty \leq x \leq \infty$
b. $x \geq 1$
c. $x \leq 3$
d. $x \geq 3$

2. Find the domain and range of the quadratic function f whose graph appears here.

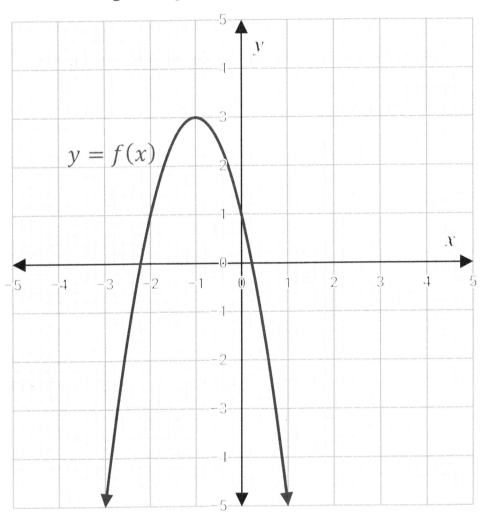

a. Domain: $-\infty \leq x \leq \infty$, Range: $y \leq 3$
b. Domain: $-\infty \leq x \leq \infty$, Range: $y \geq 3$
c. Domain: $-\infty \leq x \leq \infty$, Range: $-\infty \leq y \leq \infty$
d. Domain: $x \geq -1$, Range: $y \leq 3$

3. Find the range of the quadratic function f whose graph appears here.

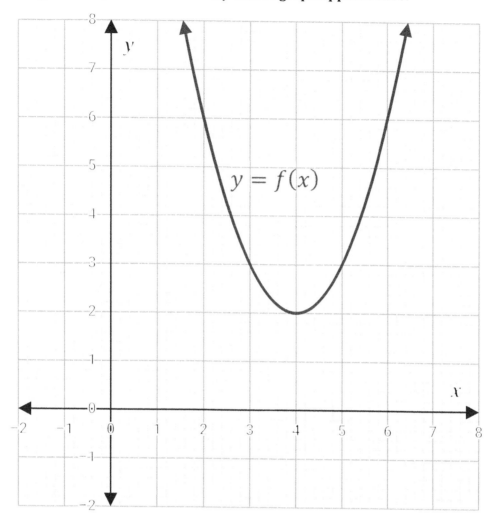

a. $y \geq 0$
b. $-\infty \leq y \leq \infty$
c. $y \geq 2$
d. $y \leq 2$

4. Find the range of the quadratic function f whose graph appears here.

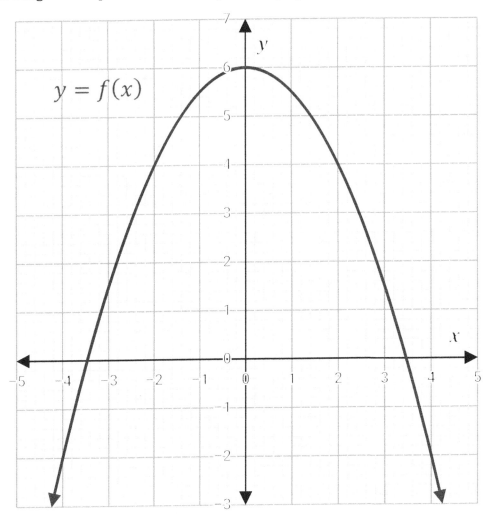

a. $y \geq 6$
b. $-\infty \leq y \leq \infty$
c. $y \leq 6$
d. $y \leq 0$

5. Find the range of the quadratic function f whose graph appears here.

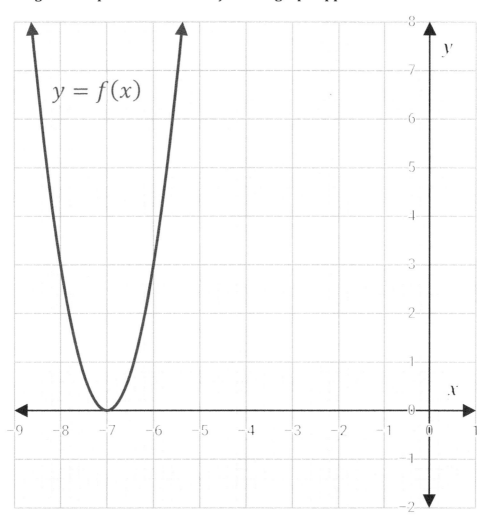

a. $-\infty \leq y \leq \infty$
b. $y \geq -7$
c. $y \geq 0$
d. $y \leq -7$

6. Find the range of the quadratic function *f* whose graph and vertex appear here.

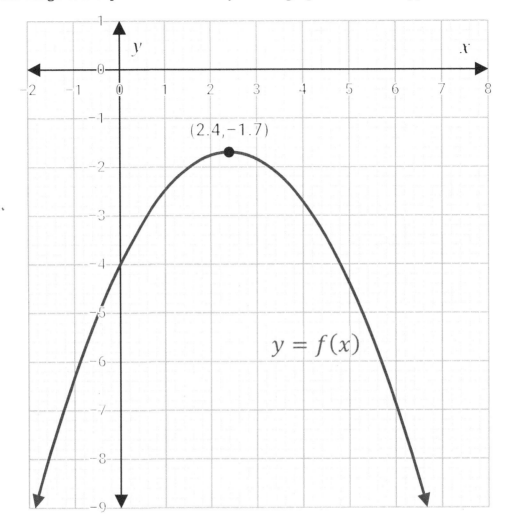

a. $y \leq 2.4$
b. $y \leq -1.7$
c. $y \geq -1.7$
d. $y \geq 2.4$

7. Find the domain of the function $f(x) = -2x^2 - 12x - 12$.

a. $x \leq 6$
b. $-\infty \leq x \leq \infty$
c. $x \leq -3$
d. $x \leq 0$

8. **Find the domain and range of the function $f(x) = x^2 - 8x + 14$.**
    a. Domain: $-\infty \leq x \leq \infty$, Range: $y \leq -2$
    b. Domain: $x \geq 4$, Range: $y \geq -2$
    c. Domain: $-\infty \leq x \leq \infty$, Range: $y \geq 4$
    d. Domain: $-\infty \leq x \leq \infty$, Range: $y \geq -2$

9. **Find the range of the function $f(x) = -x^2 - 10x - 26$.**
    a. $y \leq -1$
    b. $y \geq -1$
    c. $y \leq -5$
    d. $y \geq -5$

10. **Find the range of the function $f(x) = 6x^2 - 11$.**
    a. $y \geq 6$
    b. $y \geq 0$
    c. $y \geq -11$
    d. $y \leq -11$

11. **Find the range of the function $f(x) = -3x^2 + 42x - 147$.**
    a. $y \leq 49$
    b. $y \leq 7$
    c. $y \leq -7$
    d. $y \leq 0$

12. **Find the range of the function $f(x) = 2x^2 + 6x + 7$.**
    a. $y \geq -\frac{3}{2}$
    b. $y \leq -\frac{3}{2}$
    c. $y \leq \frac{5}{2}$
    d. $y \geq \frac{5}{2}$

13. You throw a rock straight upward, and 6 seconds later it lands at your feet. Let $s(t)$ be the height, in feet, of the rock above the ground t seconds after you throw it. Then s is a quadratic function. If the following diagram shows the graph of $s(t)$, including the vertex at $(3, 144)$, find the domain and range of the function.

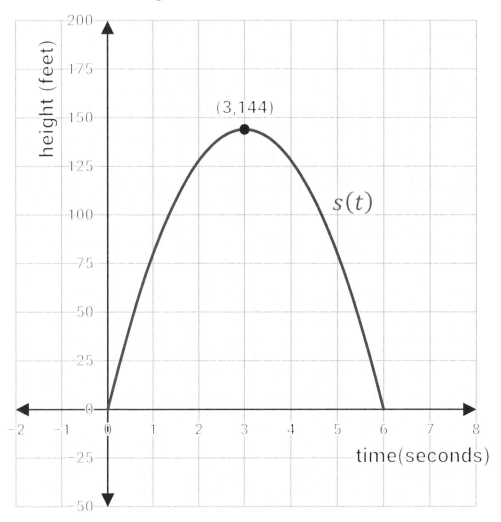

a. Domain: $0 \leq t \leq 3$, Range: $0 \leq s(t) \leq 144$
b. Domain: $0 \leq t \leq 6$, Range: $0 \leq s(t) \leq 144$
c. Domain: $0 \leq t \leq 6$, Range: $0 \leq s(t) \leq 3$
d. Domain: $0 \leq t \leq 144$, Range: $0 \leq s(t) \leq 6$

14. You manufacture widgets at a factory that can produce and sell up to 100 widgets per day. Careful research shows that a quadratic function $P(x)$, whose graph appears here with its endpoints and vertex labeled, is a good model for your profit, in dollars, when you produce x widgets per day. What is the range of $P(x)$, your range of possible daily profits?

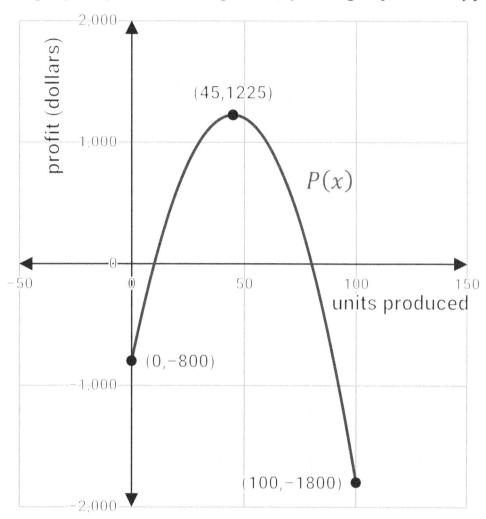

a. $-1{,}800 \leq P(x) \leq 1{,}225$
b. $-1{,}800 \leq P(x) \leq -800$
c. $10 \leq P(x) \leq 80$
d. $0 \leq P(x) \leq 100$

15. Standing on a tower, you extend your arm out and throw a rock straight upward. Four seconds later it lands on the ground below. Suppose that $s(t) = -16t^2 + 48t + 64$ is the height, in feet, of the rock above the ground t seconds after you throw it. Find the range of s.

a. $0 \leq s(t) \leq 4$
b. $0 \leq s(t) \leq 64$
c. $0 \leq s(t) \leq 100$
d. $1.5 \leq s(t) \leq 100$

8.2 Writing Equations Given Vertex and a Point

PRACTICE QUESTIONS

1. Write the equation of a quadratic function in vertex form that has a vertex at $(2, 3)$ and passes through the point $(3, 8)$.
   a. $y = 11(x - 2)^2 + 3$
   b. $y = 5(x - 2)^2 + 3$
   c. $y = 5(x^2 - 2) + 3$
   d. $y = 5(x + 2)^2 - 3$

2. Write the equation of a quadratic function in vertex form that has a vertex at $(0, 0)$ and passes through the point $(1, 3)$.
   a. $y = 2x^2$
   b. $y = \frac{1}{3}x^2$
   c. $y = x^2$
   d. $y = 3x^2$

3. Write the equation of a quadratic function in vertex form that has a vertex at $(3, -2)$ and passes through the point $(5, 2)$.
   a. $y = (x - 3)^2 - 2$
   b. $y = 4(x - 3)^2 - 2$
   c. $y = (x + 3)^2 + 2$
   d. $y = 4(x + 3)^2 + 2$

4. Write the equation of a quadratic function in vertex form that has a vertex at $(-4, 1)$ and passes through the point $(-2, 5)$.
   a. $y = 6(x + 4)^2 - 1$
   b. $y = (x^2 + 4) + 1$
   c. $y = (x + 4)^2 + 1$
   d. $y = 6(x + 4)^2 + 1$

5. Write the equation of a quadratic function in vertex form that has a vertex at $(4, 2)$ and passes through the point $(3, 3)$.
   a. $y = (x + 4)^2 - 2$
   b. $y = (x - 4)^2 + 2$
   c. $y = -(x - 4)^2 + 2$
   d. $y = 5(x - 4)^2 + 2$

6. Write the equation of a quadratic function in vertex form that has a vertex at $(4, -1)$ and passes through the point $(3, 3)$.
   a. $y = 4(x + 4)^2 - 1$
   b. $y = 4(x - 4)^2 - 1$
   c. $y = 2(x - 4)^2 - 1$
   d. $y = 4(x - 4)^2 + 1$

7. Write the equation of a quadratic function in vertex form that has a vertex at $(4, -9)$ and passes through the point $(2, -1)$.
   a. $y = 2(x^2 - 4) - 9$
   b. $y = 2(x - 4)^2 + 9$
   c. $y = 4(x - 4)^2 - 9$
   d. $y = 2(x - 4)^2 - 9$

8. Write the equation of a quadratic function in vertex form that has a vertex at $(-3, -1)$ and passes through the point $(-2, 0)$.
   a. $y = (x - 3)^2 - 1$
   b. $y = (x - 3)^2 + 1$
   c. $y = (x + 3)^2 - 1$
   d. $y = (x + 3)^2 + 1$

9. Write the equation of a quadratic function in vertex form that has a vertex at $(-1, -4)$ and passes through the point $(2, -1)$.
   a. $y = 3(x + 1)^2 + 4$
   b. $y = \frac{1}{3}(x + 1)^2 + 4$
   c. $y = 3(x + 1)^2 - 4$
   d. $y = \frac{1}{3}(x + 1)^2 - 4$

10. Write the equation of a quadratic function in vertex form that has a vertex at $(-6, 8)$ and passes through the point $(-4, 10)$.
    a. $y = \frac{1}{2}(x + 6)^2 + 8$
    b. $y = 2(x + 6)^2 + 8$
    c. $y = \frac{1}{2}x^2 + 14$
    d. $y = 2(x - 6)^2 - 8$

11. Write the equation of a quadratic function in vertex form that has a vertex at $(0, 4)$ and passes through the point $(8, -12)$.
   a. $y = -\frac{1}{4}x^2 + 4$
   b. $y = \frac{1}{4}x^2 + 4$
   c. $y = -\frac{1}{4}x^2 - 4$
   d. $y = -4x^2 + 4$

12. Write the equation of a quadratic function in vertex form that has a vertex at $(-4, 6)$ and passes through the point $(-1, 9)$.
   a. $y = 3(x + 4)^2 + 6$
   b. $y = \frac{1}{3}(x + 4)^2 + 6$
   c. $y = \frac{1}{3}(x^2 + 4) + 6$
   d. $y = \frac{1}{3}(x - 4)^2 - 6$

13. Marco is a civil engineer who is examining a model of an archway on the coordinate plane. If the highest point of the archway's graph is at the point $(0, 0)$ and it also passes through the point $(-2, -12)$, which quadratic equation best represents the archway?
   a. $y = 3x^2$
   b. $y = -3x^2$
   c. $y = -\frac{1}{3}x^2$
   d. $y = \frac{1}{3}x^2$

14. The pathway of a water jet in an outdoor fountain follows a parabolic trajectory. As the fountain's jet sprays out water, gravity pulls it back down. When graphed on the coordinate plane, this parabola has a vertex of $(-3, 0)$ and passes through the point $(-5, -4)$. Based on this information, which quadratic equation best represents the scenario?
   a. $y = -4(x + 3)^2$
   b. $y = 4(x + 3)^2$
   c. $y = (x + 3)^2$
   d. $y = -(x + 3)^2$

15. Buzzy is a structural engineer who is designing a new roller coaster. He is working on a portion of the tracks that is in the shape of a parabola. When graphed on the coordinate plane, this parabola has a vertex of $(2, 5)$ and passes through the point $(3, 7)$. Buzzy needs to write a quadratic equation in vertex form that represents this portion of the tracks. Which equation should Buzzy write?
   a. $y = 2(x - 2)^2 + 5$
   b. $y = 12(x - 2)^2 + 5$
   c. $y = -2(x - 2)^2 + 5$
   d. $y = -12(x - 2)^2 + 5$

8.3 Factoring Quadratic Equations

PRACTICE QUESTIONS

1. Solve the equation $x^2 + 9x = -8$ by factoring.
   a. $x = 9$ and $x = -8$
   b. $x = -8$ and $x = -1$
   c. $x = -9$ and $x = 1$
   d. $x = 1$ and $x = -8$

2. Solve the equation $x^2 - 6x + 8 = 0$ by factoring.
   a. $x = -2$ and $x = -4$
   b. $x = 3$ and $x = -1$
   c. $x = 2$ and $x = 4$
   d. $x = -3$ and $x = -1$

3. Solve the equation $x^2 + 2x - 15 = 0$ by factoring.
   a. $x = -3$ and $x = 5$
   b. $x = -5$ and $x = 3$
   c. $x = 2$ and $x = 8$
   d. $x = -2$ and $x = -8$

4. Solve the equation $x^2 = x + 12$ by factoring.
   a. $x = 12$ and $x = 1$
   b. $x = -4$ and $x = 3$
   c. $x = -6$ and $x = 2$
   d. $x = -3$ and $x = 4$

5. Solve the equation $2x^2 + 5x = -3$ by factoring.
   a. $x = \frac{2}{3}$ and $x = -3$
   b. $x = -\frac{3}{2}$ and $x = -1$
   c. $x = \frac{2}{3}$ and $x = 1$
   d. $x = 5$ and $x = 3$

6. Solve the equation $x^2 = 15x - 56$ by factoring.
   a. $x = 7$ and $x = 8$
   b. $x = -3$ and $x = -5$
   c. $x = -7$ and $x = -8$
   d. $x = 3$ and $x = 5$

7. Solve the equation $2x^2 + x = 21$ by factoring.
   a. $x = -\frac{7}{2}$ and $x = 3$
   b. $x = -14$ and $x = 3$
   c. $x = 2$ and $x = 21$
   d. $x = -3$ and $x = 7$

8. Solve the equation $x^2 = 7x + 18$ by factoring.
   a. $x = -3$ and $x = 6$
   b. $x = -9$ and $x = 2$
   c. $x = -2$ and $x = 9$
   d. $x = -6$ and $x = 3$

9. Solve the equation $2(x^2 - 6x + 8) = 0$ by factoring.
   a. $x = 2$ and $x = 4$
   b. $x = \frac{1}{2}$ and $x = 4$
   c. $x = 8$ and $x = -4$
   d. $x = 2$ and $x = -8$

10. Solve the equation $x^2 + 6x + 8 = 0$ by factoring.
   a. $x = 3$
   b. $x = 6$ and $x = 9$
   c. $x = 2$ and $x = 4$
   d. $x = -4$ and $x = -2$

11. Solve the equation $9x^2 = 64$ by factoring.
   a. $x = -\frac{3}{8}$ and $x = \frac{3}{8}$
   b. $x = -3$ and $x = 8$
   c. $x = -8$ and $x = 3$
   d. $x = -\frac{8}{3}$ and $x = \frac{8}{3}$

12. Solve the equation $x^2 - 49 = 0$ by factoring.
   a. $x = 1$ and $x = -49$
   b. $x = -7$ and $x = 7$
   c. $x = 0$
   d. $x = -7$ and $x = 49$

13. A plot of land has a width of w feet and a length of $2w + 3$ feet. If the plot of land has an area of 90 square feet, what is the width of the land?
   a. 5 feet
   b. 15 feet
   c. 6 feet
   d. $\frac{15}{2}$ feet

14. Maddie is creating a garden in her backyard. The area of her garden is 24 square meters. The length of the garden is 2 meters more than its width. What is the length of the garden?
   a. 10 meters
   b. 8 meters
   c. 6 meters
   d. 4 meters

15. The product of two numbers is 45, and their sum is 14. Which quadratic equation represents this scenario where s is one of the numbers?
   a. $s(45 + s) = 14$
   b. $s(45 - s) = 14$
   c. $s(14 + s) = 45$
   d. $s(14 - s) = 45$

8.4 Writing Quadratic Functions Given Solutions and Graphs

PRACTICE QUESTIONS

1. Write an equation in standard form for the quadratic function shown below, with roots at $x = -3$ and $x = 4$.

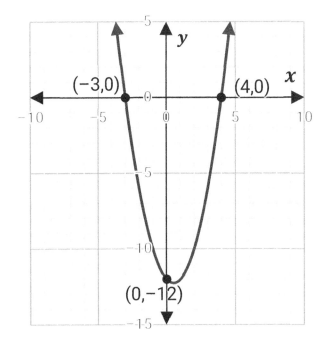

a. $y = x^2 - 7x - 7$
b. $y = x^2 + 7x + 7$
c. $y = -x^2 + x + 12$
d. $y = x^2 - x - 12$

2. Write an equation in standard form for the quadratic function shown below, with roots at $x = 2$ and $x = 6$.

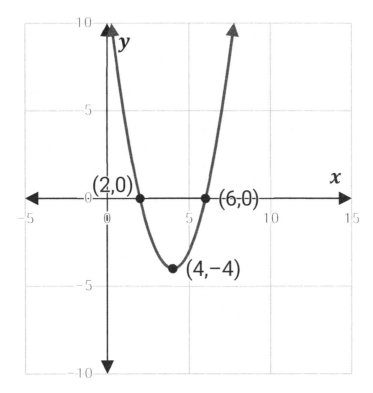

a. $y = x^2 - 8x + 12$
b. $y = x^2 + 8x + 12$
c. $y = x^2 + 8x - 8$
d. $y = x^2 - 12x + 8$

3. Write an equation in standard form for the quadratic function shown below, with roots at $x = -1$ and $x = 5$.

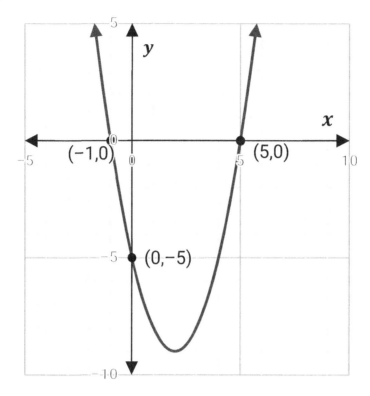

a. $y = x^2 - 6x + 5$
b. $y = x^2 + x - 5$
c. $y = x^2 - 5x - 5$
d. $y = x^2 - 4x - 5$

4. Write an equation in standard form for the quadratic function shown below.

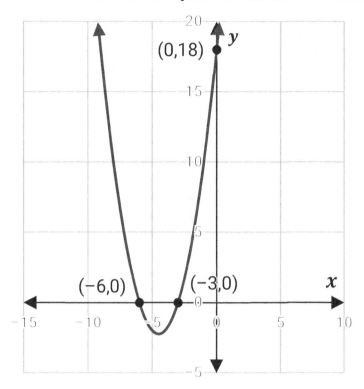

a. $y = x^2 + 9x + 18$
b. $y = x^2 - 9x + 18$
c. $y = x^2 + 9x - 9$
d. $y = x^2 - 9x + 9$

5. Write an equation in standard form for the quadratic function shown below.

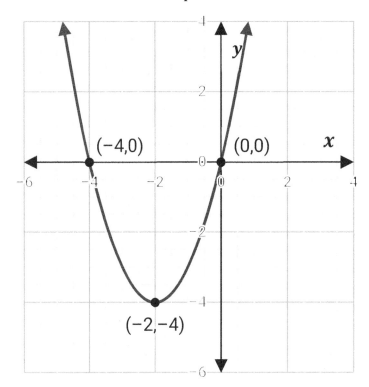

a. $y = x^2 + 4$
b. $y = x^2 - 4$
c. $y = x^2 + 4x$
d. $y = x^2 - 4x$

6. Write an equation in standard form for the quadratic function shown below.

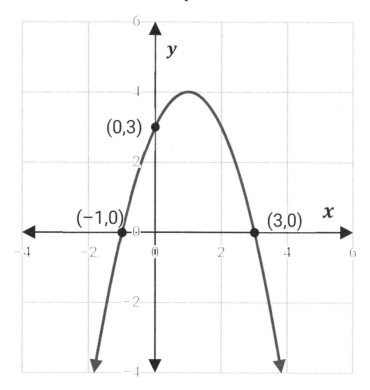

a. $y = x^2 - 2x - 3$
b. $y = -x^2 + 2x + 3$
c. $y = -x^2 - 2x - 3$
d. $y = -x^2 - 4x + 3$

7. Write an equation in standard form for the quadratic function shown below.

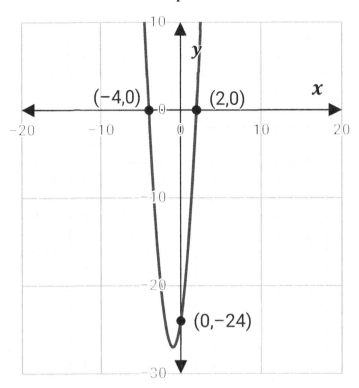

a. $y = x^2 + 2x - 8$
b. $y = 3x^2 + 6x - 24$
c. $y = 3x^2 + 2x - 8$
d. $y = 3x^2 + 5x - 5$

8. Write an equation in standard form for the quadratic function shown below.

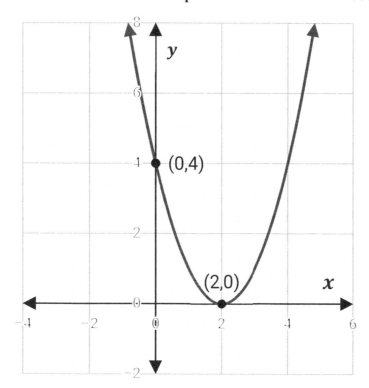

a. $y = x^2 - 4x + 4$
b. $y = 4x^2 - 16x + 16$
c. $y = x^2 + 4x - 4$
d. $y = 4x^2 + 16x - 16$

9. Write an equation in standard form for the quadratic function shown below.

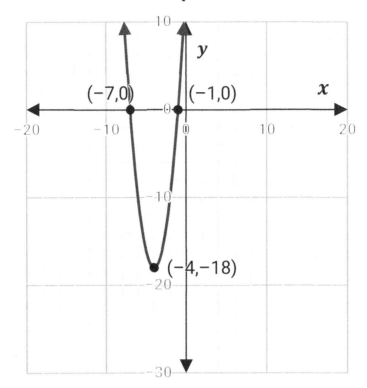

a. $y = 2x^2 - 16x + 14$
b. $y = x^2 - 8x + 7$
c. $y = x^2 + 8x + 7$
d. $y = 2x^2 + 16x + 14$

10. Write an equation in standard form for the quadratic function shown below.

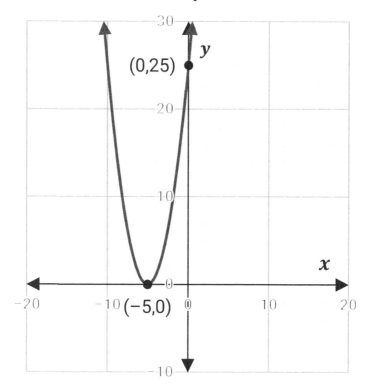

a. $y = x^2 - 10x - 10$
b. $y = x^2 + 10x + 10$
c. $y = x^2 + 10x + 25$
d. $y = x^2 - 10x + 25$

11. Write an equation in standard form for the quadratic function shown below.

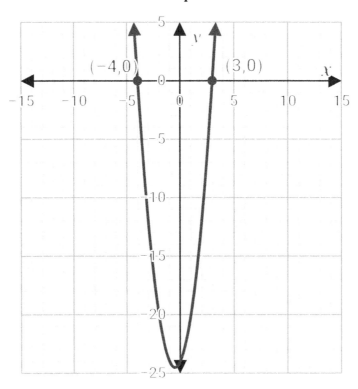

a. $y = -x^2 + 2x + 24$
b. $y = x^2 + x - 12$
c. $y = 2x^2 + 2x - 24$
d. $y = 2x^2 + 2x - 12$

12. Write an equation in standard form for the quadratic function shown below.

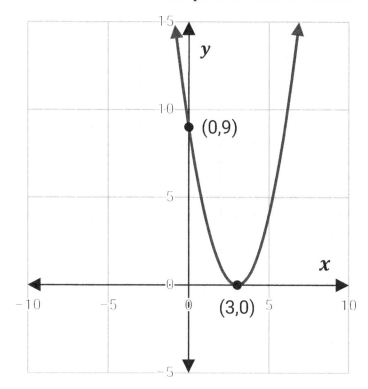

a. $y = -x^2 - 3x + 6$
b. $y = x^2 + 3x - 6$
c. $y = x^2 - 6x + 9$
d. $y = -x^2 + 6x - 9$

13. Marc bounces a basketball and watches as it rebounds back up into the air. As he observes the basketball, he notices that its path takes the shape of a parabola. Marc sketches the basketball's path on a coordinate plane, as shown below. Based on the graph, what is the equation of this function?

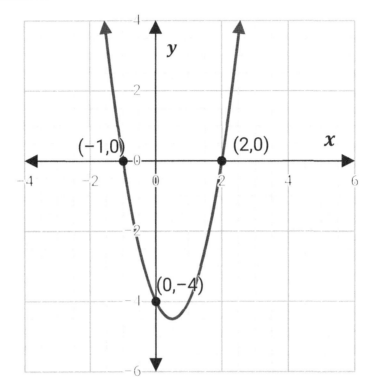

a. $y = -x^2 - x - 2$
b. $y = -2x^2 + 2x + 4$
c. $y = x^2 + x + 1$
d. $y = 2x^2 - 2x - 2$

14. Alen is a graphic designer who is overseeing the creation of a new logo. Part of the logo is in the shape of a parabola, as shown on the coordinate plane below. Alen needs to write an equation for this portion of the logo. Which quadratic equation, in standard form, represents Alen's logo?

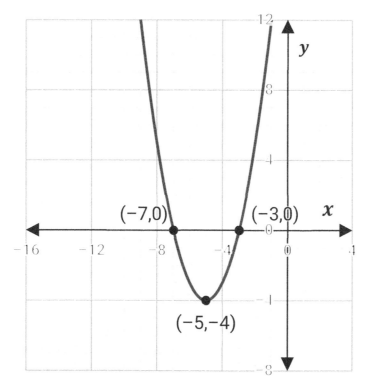

a. $y = x^2 + 10x + 21$
b. $y = -x^2 - 10x - 21$
c. $y = x^2 + 10x + 10$
d. $y = -x^2 - 4x - 10$

15. A logo is made up of two parabolic curves, one of which is represented on the coordinate plane below. What is the quadratic equation, written in standard form, of this function?

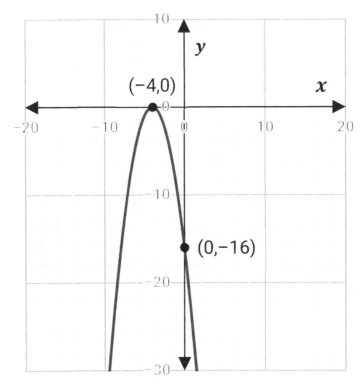

a. $y = x^2 + 16x + 8$
b. $y = -x^2 - 16x - 8$
c. $y = x^2 + 8x + 16$
d. $y = -x^2 - 8x - 16$

8.5 Solving Quadratic Equations Using Square Roots

PRACTICE QUESTIONS

1. Solve the quadratic equation by taking square roots.
$$x^2 - 9 = 7$$
   a. $x = 4$ or $x = -4$
   b. $x = 2$ or $x = -2$
   c. $x = 4$
   d. $x = 2$

2. Solve the quadratic equation by taking square roots.
$$(x - 3)^2 = 36$$
   a. $x = 39$ or $x = -39$
   b. $x = 9$ or $x = -3$
   c. $x = 3$ or $x = -3$
   d. $x = 9$ or $x = -9$

3. Solve the quadratic equation by taking square roots.
$$\frac{1}{2}x^2 = 50$$
   a. $x = 5$ or $x = -5$
   b. $x = 25$ or $x = -25$
   c. $x = 10$ or $x = -10$
   d. $x = 100$ or $x = -100$

4. Solve the quadratic equation by taking square roots.
$$(x + 7)^2 - 64 = 0$$
   a. $x = 15$ or $x = -15$
   b. $x = 1$ or $x = -1$
   c. $x = 1$ or $x = -15$
   d. $x = 57$ or $x = -57$

5. Solve the quadratic equation by taking square roots.
$$8x^2 - 8 = 192$$
   a. $x = 20$ or $x = -20$
   b. $x = 5$
   c. $x = -20$
   d. $x = 5$ or $x = -5$

6. Solve the quadratic equation by taking square roots:
$$(x-8)^2 + 9 = 130$$

   a. $x = 19$
   b. $x = 11$
   c. $x = 19$ or $x = -3$
   d. $x = 11$ or $x = -11$

7. Solve the quadratic equation by taking square roots.
$$4x^2 - 2 = 14$$

   a. $x = 4$ or $x = -4$
   b. $x = 2$ or $x = -2$
   c. $x = 2$ or $x = -4$
   d. $x = -2$ or $x = 4$

8. Solve the quadratic equation by taking square roots.
$$2x^2 + 4 = 22$$

   a. $x = 3$ or $x = -9$
   b. $x = 18$ or $x = -18$
   c. $x = 9$ or $x = -9$
   d. $x = 3$ or $x = -3$

9. Solve the quadratic equation by taking square roots.
$$2(x-1)^2 - 3 = 47$$

   a. $x = 6$ or $x = -4$
   b. $x = 4$ or $x = -6$
   c. $x = 4$ or $x = -4$
   d. $x = 5$ or $x = -5$

10. Solve the quadratic equation by taking square roots.
$$4x^2 + 24 = 600$$

    a. $x = 12$ or $x = -12$
    b. $x = 144$ or $x = -144$
    c. $x = 12$ or $x = -144$
    d. $x = 144$ or $x = -12$

11. Solve the quadratic equation by taking square roots.
$$9(2x-3)^2 + 6 = 447$$

    a. $x = 5$ or $x = -5$
    b. $x = 2$ or $x = -2$
    c. $x = 5$ or $x = -2$
    d. $x = 2$ or $x = -5$

12. Solve the quadratic equation by taking square roots.
$$x^2 + 6 = 26$$
   a. $x = 2$ or $x = -2$
   b. $x = 5$ or $x = -5$
   c. $x = 20$ or $x = -20$
   d. $x = 2\sqrt{5}$ or $x = -2\sqrt{5}$

13. The length of a rectangle equals twice its width, and the area is 162 square inches. What is the length of the rectangle?
   a. 2 inches
   b. 81 inches
   c. 18 inches
   d. 9 inches

14. Tim and Mary Kate are building a square patio. They have enough building materials for the patio to have an area of 169 square feet. The length of one of the sides of the patio is represented by the expression $x + 5$. Write a quadratic equation that represents this scenario and solve it by taking square roots. What is the perimeter of the patio?
   a. 13 feet
   b. 52 feet
   c. 169 feet
   d. 8 feet

15. A pen is dropped out of a window that is 400 feet above the ground. The speed at which the pen will hit the ground can be calculated using the formula $s^2 = 64h$, where s stands for the speed of the falling object, and h stands for the distance the object falls. Based on this information, solve the equation by taking square roots to find the speed of the pen when it hit the ground.
   a. 25,600 feet per second
   b. 160 feet per second
   c. −160 feet per second
   d. 400 feet per second

8.6 Completing the Square

PRACTICE QUESTIONS

1. Solve the quadratic equation by completing the square: $x^2 - 12x + 11 = 0$.
   a. $x = 6$ or $x = 5$
   b. $x = 11$ or $x = 1$
   c. $x = -1$ or $x = 11$
   d. $x = 36$ or $x = -6$

2. Solve the quadratic equation by completing the square: $x^2 + 6x + 8 = 0$.
   a. $x = -2$ or $x = -4$
   b. $x = 2$ or $x = 4$
   c. $x = 3$ or $x = -8$
   d. $x = 3$ or $x = 9$

3. Solve the quadratic equation by completing the square: $x^2 + 8x + 7 = 0$.
   a. $x = 12$ or $x = -12$
   b. $x = 4$ or $x = 16$
   c. $x = 1$ or $x = -8$
   d. $x = -1$ or $x = -7$

4. Solve the quadratic equation by completing the square: $12x^2 - 24x - 36 = 0$.
   a. $x = -3$ or $x = -1$
   b. $x = 2$ or $x = 3$
   c. $x = 1$ or $x = -1$
   d. $x = 3$ or $x = -1$

5. Solve the quadratic equation by completing the square: $2x^2 - 20x + 32 = 0$.
   a. $x = 5$ or $x = 3$
   b. $x = 8$ or $x = 2$
   c. $x = -5$ or $x = 5$
   d. $x = 4$ or $x = -2$

6. Solve the quadratic equation by completing the square: $6x^2 + 12x - 48 = 0$.
   a. $x = 1$ or $x = -3$
   b. $x = 2$ or $x = -4$
   c. $x = 1$ or $x = 4$
   d. $x = 2$ or $x = -3$

7. Solve the quadratic equation by completing the square: $5x^2 + 20x - 60 = 0$.
   a. $x = 2$ or $x = -4$
   b. $x = 4$ or $x = 6$
   c. $x = 2$ or $x = -6$
   d. $x = -2$ or $x = -4$

8. Solve the quadratic equation by completing the square: $4x^2 - 12x - 16 = 0$.
   a. $x = 4$ or $x = -1$
   b. $x = 25$ or $x = 4$
   c. $x = 2$ or $x = 5$
   d. $x = 4$ or $x = 2$

9. Solve the quadratic equation by completing the square: $4x^2 + 20x + 25 = 0$.
   a. $x = -\frac{5}{2}$ or $x = 0$
   b. $x = \frac{5}{2}$ or $x = 0$
   c. $x = -\frac{5}{2}$
   d. $x = \frac{5}{2}$

10. Solve the quadratic equation by completing the square: $3x^2 + 8x - 3 = 0$.
   a. $x = \frac{1}{6}$ or $x = 3$
   b. $x = 10$ or $x = 6$
   c. $x = 6$ or $x = 3$
   d. $x = -3$ or $x = \frac{1}{3}$

11. Solve the quadratic equation by completing the square: $3x^2 - 5x + 2 = 0$.
   a. $x = 6$ or $x = \frac{5}{6}$
   b. $x = 1$ or $x = \frac{2}{3}$
   c. $x = 3$ or $x = 1$
   d. $x = \frac{2}{3}$ or $x = 36$

12. Solve the quadratic equation by completing the square: $x^2 + 6x - 3 = 0$.
   a. $x = 2 \pm 3\sqrt{2}$
   b. $x = 2$ or $x = -3$
   c. $x = 5$ or $x = 1$
   d. $x = -3 \pm 2\sqrt{3}$

13. A ride at an amusement park takes passengers to the top of a tower and then drops them 57 feet. The quadratic equation that represents this ride is $-16t^2 + 64t + 57 = 0$, where t stands for the time, in seconds, it takes for riders to reach ground level. Solve the equation for t by completing the square to figure out how many seconds it will take for riders to reach ground level.

    a. $4\frac{3}{4}$ seconds
    b. $2\frac{3}{4}$ seconds
    c. 11 seconds
    d. 4 seconds

14. Jill is framing a picture. The length of the frame is 6 inches longer than its width, and the picture and frame has a total area of 520 square inches. Given this information, write a quadratic equation to represent this scenario and solve it by completing the square to identify the dimensions of the frame.

    a. The width of the frame is 20 inches, and the length is 26 inches.
    b. The width of the frame is 22 inches, and the length is 28 inches.
    c. The width of the frame is 26 inches, and the length is 32 inches.
    d. The width of the frame is 20 inches, and the length is 26 inches.

15. Miguel throws a football straight up into the air, from 3 feet above the ground, with a velocity of 14 feet per second. The quadratic equation that represents this scenario is $5t^2 - 14t - 3 = 0$, where t stands for time in seconds. Solve the equation for t by completing the square to figure out how many seconds it will take for the football to hit the ground.

    a. 7 seconds
    b. 15 seconds
    c. 3 seconds
    d. $\frac{1}{5}$ second

8.7 Converting Between Standard and Vertex Forms

Practice Questions

1. Convert the quadratic equation from vertex form to standard form.
$$y = 3(x - 1)^2 + 2$$
   a. $y = 3x^2 - 6x + 3$
   b. $y = 3x^2 - 3x - 6$
   c. $y = 3x^2 - 2x + 3$
   d. $y = 3x^2 - 6x + 5$

2. Convert the quadratic equation from standard form to vertex form.
$$y = 2x^2 - 4x + 5$$
   a. $y = 2(x - 3)^2 + 1$
   b. $y = 2(x - 1)^2 + 3$
   c. $y = 2(x + 1)^2 - 3$
   d. $y = 2(x + 3)^2 - 1$

3. Convert the quadratic equation from vertex form to standard form.
$$y = (x + 3)^2 - 1$$
   a. $y = x^2 + 6x + 9$
   b. $y = x^2 + 6x + 8$
   c. $y = x^2 + 9x + 8$
   d. $y = x + 8$

4. Convert the quadratic equation from standard form to vertex form.
$$y = 2x^2 - 8x + 9$$
   a. $y = 2(x - 2)^2 + 1$
   b. $y = 2(x + 2)^2 + 1$
   c. $y = 2(x - 1)^2 + 2$
   d. $y = 2(x - 4)^2 + 1$

5. Convert the quadratic equation from vertex form to standard form.
$$y = 2(x + 2)^2 + 7$$
   a. $y = 2x^2 + 4x + 15$
   b. $y = 2x^2 + 4x + 11$
   c. $y = 2x^2 + 8x + 15$
   d. $y = 2x^2 + 8x + 1$

6. Convert the quadratic equation from standard form to vertex form.
$$y = x^2 - 4x - 3$$
   a. $y = (x - 2)^2 - 7$
   b. $y = (x + 7)^2 - 2$
   c. $y = (x + 2)^2 + 7$
   d. $y = (x - 7)^2 + 2$

7. Convert the quadratic equation from vertex form to standard form.
$$y = 4(x + 5)^2 - 25$$
   a. $y = 4x^2 + 10x$
   b. $y = 4x^2 + 40x + 75$
   c. $y = 4x^2 + 10x + 75$
   d. $y = 4x^2 + 40x$

8. Convert the quadratic equation from standard form to vertex form.
$$y = x^2 + 8x + 20$$
   a. $y = (x - 4)^2 - 4$
   b. $y = (x + 4)^2 - 4$
   c. $y = (x - 4)^2 + 4$
   d. $y = (x + 4)^2 + 4$

9. Convert the quadratic equation from vertex form to standard form.
$$y = -(x - 6)^2$$
   a. $y = -x^2 + 12x - 36$
   b. $y = x^2 - 12x + 36$
   c. $y = -x^2 + 12x - 12$
   d. $y = x^2 - 12x + 12$

10. Convert the quadratic equation from standard form to vertex form.
$$y = -x^2 + 2x - 2$$
   a. $y = -(x - 1)^2 - 1$
   b. $y = -(x + 1)^2 - 1$
   c. $y = (x - 1)^2 - 1$
   d. $y = -(x - 1)^2 + 1$

11. Convert the quadratic equation from vertex form to standard form.
$$y = \frac{1}{2}(x-4)^2 - 3$$

   a. $y = \frac{1}{2}x^2 - 8x + 5$
   b. $y = \frac{1}{2}x^2 - 8x + 13$
   c. $y = \frac{1}{2}x^2 - 4x - 3$
   d. $y = \frac{1}{2}x^2 - 4x + 5$

12. Convert the quadratic equation from standard form to vertex form.
$$y = x^2 + 2x - 6$$

   a. $y = (x-1)^2 + 8$
   b. $y = (x+1)^2 - 8$
   c. $y = (x+1)^2 - 7$
   d. $y = (x-1)^2 - 7$

13. The arc of a football for a particular throw can be modeled by the quadratic equation $y = -\frac{1}{100}(x-20)^2 + 8$. What is the standard form of this quadratic equation?

   a. $y = \frac{1}{100}x^2 - \frac{2}{5}x - 12$
   b. $y = -\frac{1}{100}x^2 + \frac{2}{5}x + 12$
   c. $y = -\frac{1}{100}x^2 + \frac{2}{5}x + 4$
   d. $y = \frac{1}{100}x^2 + \frac{2}{5}x - 4$

14. The equation $y = x^2 - 12x + 45$ models the number of records, y, sold in a record store x days after a band played at a publicity event. What is the vertex form of this quadratic equation?

   a. $y = (x-6)^2 + 9$
   b. $y = (x-9)^2 + 6$
   c. $y = (x+6)^2 - 9$
   d. $y = (x+9)^2 - 6$

15. Mr. Dawson's physics class is launching a rocket from a platform. The height of the rocket in feet x seconds after the launch is represented by the equation $y = -5(x-4)^2 + 200$. What is the standard form of this quadratic equation?

   a. $y = -5x^2 - 8x + 216$
   b. $y = -5x^2 + 40x + 120$
   c. $y = x^2 - 8x + 16$
   d. $y = -5x^2 - 40x - 120$

8.8 The Quadratic Formula

PRACTICE QUESTIONS

1. Solve using the quadratic formula: $x^2 + x - 6 = 0$.
   a. $x = 3$ and $x = -2$
   b. $x = 2$ and $x = -3$
   c. $x = 6$ and $x = 4$
   d. $x = -6$ and $x = -4$

2. Solve using the quadratic formula: $x^2 + 4x + 3 = 0$.
   a. $x = -2$ and $x = -6$
   b. $x = 0$ and $x = -4$
   c. $x = -1$ and $x = -3$
   d. $x = 1$ and $x = 3$

3. Solve using the quadratic formula: $x^2 - 6x - 16 = 0$.
   a. $x = 16$ and $x = -4$
   b. $x = 6$ and $x = 0$
   c. $x = 8$ and $x = -2$
   d. $x = 8$ and $x = 2$

4. Solve using the quadratic formula: $x^2 - 8x + 12 = 0$.
   a. $x = 12$ and $x = -8$
   b. $x = -4$ and $x = -12$
   c. $x = -6$ and $x = -2$
   d. $x = 6$ and $x = 2$

5. Solve using the quadratic formula: $2x^2 + 2x - 12 = 0$.
   a. $x = 2$ and $x = -3$
   b. $x = -12$ and $x = 8$
   c. $x = 24.5$ and $x = -25.5$
   d. $x = -8$ and $x = 12$

6. Solve using the quadratic formula: $5x^2 - 80 = 0$.
   a. $x = 0$ and $x = 40$
   b. $x = 5$ and $x = 80$
   c. $x = 40$ and $x = -40$
   d. $x = 4$ and $x = -4$

7. Solve using the quadratic formula: $4x^2 + 2x - 42 = 0$.
   a. $x = 2$ and $x = -42$
   b. $x = 7$ and $x = 4$
   c. $x = 3$ and $x = -4$
   d. $x = 3$ and $x = -\frac{7}{2}$

8. Solve using the quadratic formula: $2x^2 - 14x + 24 = 0$.
   a. $x = 4$ and $x = 3$
   b. $x = -4$ and $x = -3$
   c. $x = 9$ and $x = 5$
   d. $x = -9$ and $x = -5$

9. Solve using the quadratic formula: $3x^2 - x - 4 = 0$.
   a. $x = \frac{3}{4}$ and $x = \frac{4}{3}$
   b. $x = \frac{4}{3}$ and $x = -1$
   c. $x = -\frac{4}{3}$ and $x = -1$
   d. $x = 6$ and $x = 8$

10. Solve using the quadratic formula: $8x^2 + 14x + 3 = 0$.
   a. $x = \frac{1}{2}$ and $x = \frac{1}{4}$
   b. $x = -12$ and $x = -2$
   c. $x = -24$ and $x = -4$
   d. $x = -\frac{1}{4}$ and $x = -\frac{3}{2}$

11. Solve using the quadratic formula: $4x^2 + 7x - 15 = 0$.
   a. $x = 7$ and $x = 4$
   b. $x = 5$ and $x = -12$
   c. $x = \frac{5}{4}$ and $x = -3$
   d. $x = \frac{4}{5}$ and $x = 3$

12. Solve using the quadratic formula: $5x^2 + 50x + 125 = 0$.
   a. $x = 0$
   b. $x = 0$ or $x = -5$
   c. $x = -5$
   d. $x = 5$ or $x = -5$

13. The length of a rectangular flag is 12 inches more than its width, and the area is 325 square inches. Write a quadratic equation that represents this scenario and solve it by using the quadratic formula to identify the length and width of the flag.
    a. The width of the flag is 14 inches, and the length is 26 inches.
    b. The width of the flag is 13 inches, and the length is 25 inches.
    c. The width of the flag is 12 inches, and the length is 24 inches.
    d. The width of the flag is 16 inches, and the length is 28 inches.

14. A pack of gum is thrown out of a window. The speed at which the pack of gum will hit the ground can be calculated using the formula $-2v^2 + 5v + 900 = 0$, where v stands for the velocity of the falling object. Based on this information, solve the equation by using the quadratic formula to find the speed of the pack of gum when it hits the ground in feet per second. (Hint: speed is the absolute value of velocity.)
    a. −20 feet per second
    b. 22.5 feet per second
    c. 85 feet per second
    d. 127.5 feet per second

15. Nicole is installing a concrete slab in the shape of a rectangle in her backyard with an area of 540 square feet. The length of the rectangle is represented by the expression $x - 4$. The width of the rectangle is represented by the expression $x + 8$. Write a quadratic equation that represents this scenario and solve it by using the quadratic formula to find the perimeter of the concrete slab.
    a. 22 feet
    b. 36 feet
    c. 60 feet
    d. 96 feet

8.9 The Discriminant

PRACTICE QUESTIONS

1. Calculate the discriminant of the quadratic equation below. Use this information to determine how many times the graph of this function touches the x-axis.
$$x^2 + 9x + 14 = 0$$
   a. $D = -25$. The graph touches the x-axis 0 times.
   b. $D = 25$. The graph touches the x-axis 1 time.
   c. $D = 25$. The graph touches the x-axis 2 times.
   d. $D = -25$. The graph touches the x-axis 2 times.

2. Calculate the discriminant of the quadratic equation below. Use this information to determine how many times the graph of this function will touch the x-axis.
$$x^2 - 8x + 16 = 0$$
   a. $D = 0$. The graph touches the x-axis 0 times.
   b. $D = 0$. The graph touches the x-axis 1 time.
   c. $D = 64$. The graph touches the x-axis 2 times.
   d. $D = 64$. The graph touches the x-axis 1 time.

3. Calculate the discriminant of the quadratic equation below. Use this information to determine how many times the graph of this function will touch the x-axis.
$$x^2 - 4x + 8 = 0$$
   a. $D = 16$. The graph touches the x-axis 2 times.
   b. $D = -16$. The graph touches the x-axis 1 time.
   c. $D = 16$. The graph touches the x-axis 0 times.
   d. $D = -16$. The graph touches the x-axis 0 times.

4. Calculate the discriminant of the quadratic equation below. Use this information to determine how many times the graph of this function will touch the x-axis.
$$3x^2 - 12x + 12 = 0$$
   a. $D = 0$. The graph touches the x-axis 0 times.
   b. $D = 0$. The graph touches the x-axis 1 time.
   c. $D = 144$. The graph touches the x-axis 1 time.
   d. $D = 288$. The graph touches the x-axis 2 times.

5. Calculate the discriminant of the quadratic equation below. Use this information to determine how many times the graph of this function will touch the x-axis.
$$5x^2 - x - 1 = 0$$
   a. $D = 19$. The graph touches the x-axis 2 times.
   b. $D = -19$. The graph touches the x-axis 0 times.
   c. $D = -21$. The graph touches the x-axis 0 times.
   d. $D = 21$. The graph touches the x-axis 2 times.

6. Calculate the discriminant of the quadratic equation below. Use this information to determine how many times the graph of this function will touch the x-axis.

$$x^2 - 6x + 15 = 0$$

a. $D = -24$. The graph touches the x-axis 0 times.
b. $D = 24$. The graph touches the x-axis 2 times.
c. $D = -48$. The graph touches the x-axis 0 times.
d. $D = 72$. The graph touches the x-axis 2 times.

7. Calculate the discriminant of the quadratic equation below. Use this information to determine how many times the graph of this function will touch the x-axis.

$$2x^2 - 8x - 13 = 0$$

a. $D = 168$. The graph touches the x-axis 2 times.
b. $D = -40$. The graph touches the x-axis 0 times.
c. $D = 40$. The graph touches the x-axis 2 times.
d. $D = 120$. The graph touches the x-axis 2 times.

8. Calculate the discriminant of the quadratic equation below. Use this information to determine how many times the graph of this function will touch the x-axis.

$$-3x^2 - 10x - 9 = 0$$

a. $D = -8$. The graph touches the x-axis 0 times.
b. $D = 8$. The graph touches the x-axis 2 times.
c. $D = 208$. The graph touches the x-axis 2 times.
d. $D = -88$. The graph touches the x-axis 0 times.

9. Calculate the discriminant of the quadratic equation below. Use this information to determine how many times the graph of this function will touch the x-axis.

$$x^2 - 6x + 9 = 0$$

a. $D = -24$. The graph touches the x-axis 0 times.
b. $D = 72$. The graph touches the x-axis 2 times.
c. $D = 0$. The graph touches the x-axis 1 time.
d. $D = 0$. The graph touches the x-axis 0 times.

10. Calculate the discriminant of the quadratic equation below. Use this information to determine how many times the graph of this function will touch the x-axis.

$$-3x^2 + 5x - 11 = 0$$

a. $D = 107$. The graph touches the x-axis 2 times.
b. $D = -107$. The graph touches the x-axis 0 times.
c. $D = -122$. The graph touches the x-axis 0 times.
d. $D = 122$. The graph touches the x-axis 2 times.

11. Calculate the discriminant of the quadratic equation below. Use this information to determine how many times the graph of this function will touch the x-axis.

$$-5x^2 - 16x - 12 = 0$$

a. $D = 0$. The graph touches the x-axis 1 time.
b. $D = -16$. The graph touches the x-axis 0 times.
c. $D = 16$. The graph touches the x-axis 2 times.
d. $D = -208$. The graph touches the x-axis 2 times.

12. Calculate the discriminant of the quadratic equation below. Use this information to determine how many times the graph of this function will touch the x-axis.

$$x^2 - 12x + 36 = 0$$

a. $D = 120$. The graph touches the x-axis 2 times.
b. $D = -120$. The graph touches the x-axis 0 times.
c. $D = 0$. The graph touches the x-axis 1 time.
d. $D = 288$. The graph touches the x-axis 2 times.

13. Is it possible for the product of a number and 7 less than that number to be 256?

a. Yes, because the discriminant is positive.
b. Yes, because the discriminant is zero.
c. No, because the discriminant is negative.
d. Solving for the discriminant will not give enough information to answer this question.

14. Is it possible for the product of 3 times a number and 4 more than that number to equal –117?

a. Yes, because the discriminant is positive.
b. Yes, because the discriminant is zero.
c. No, because the discriminant is negative.
d. Solving for the discriminant will not give enough information to answer this question.

15. Elizabeth is in charge of finances for her company and created an equation to represent the company's profits for the year. The equation she comes up with is $f(x) = 7x^2 - 20x + 954$. Will the company be able to turn a profit of $10,000 this year?

a. Yes, because the discriminant is positive.
b. Yes, because the discriminant is zero.
c. No, because the discriminant is negative.
d. Solving for the discriminant will not give enough information to answer this question.

8.10 Graphing Quadratic Functions

PRACTICE QUESTIONS

1. Which graph matches the equation $y = x^2 - 2x - 3$?

a.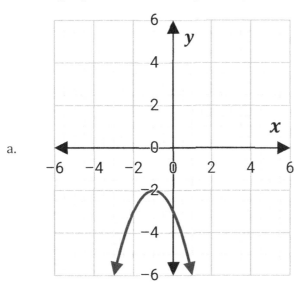

c.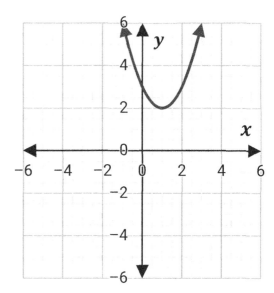

b.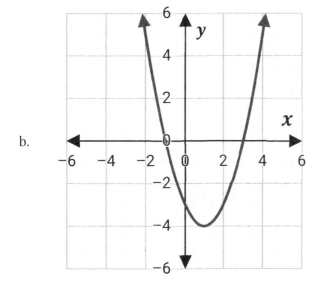

d.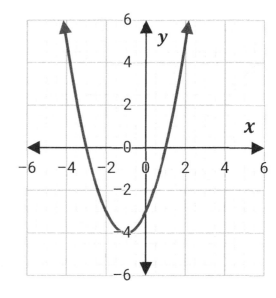

2. Which graph matches the equation $y = x^2 - 4x - 5$?

a.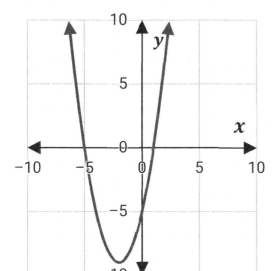

c.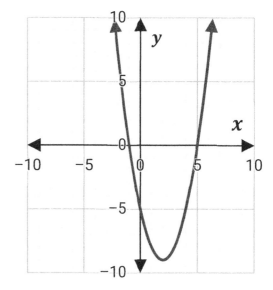

b.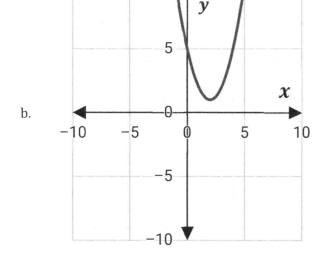

d.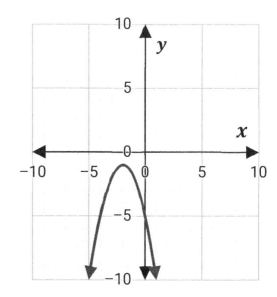

3. Which graph matches the equation $y = 3x^2 - 12x + 9$?

a.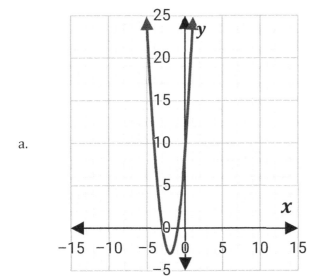

c.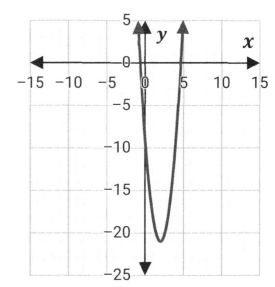

b.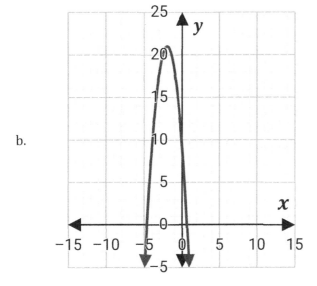

d.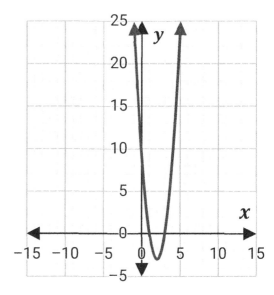

4. Which graph matches the equation $y = x^2 - 10x + 9$?

a.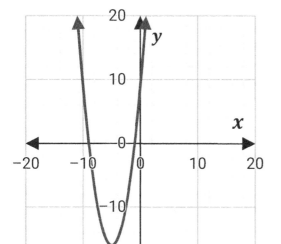

c.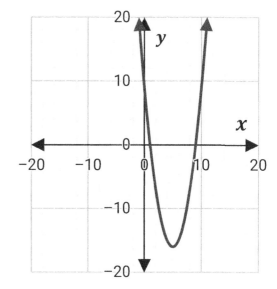

b.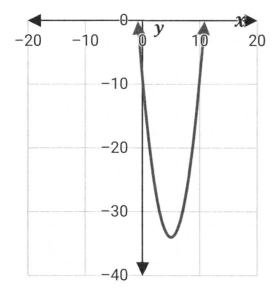

d.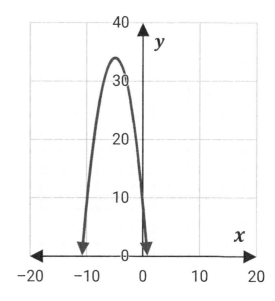

5. Which graph matches the equation $y = x^2 - 1$?

a.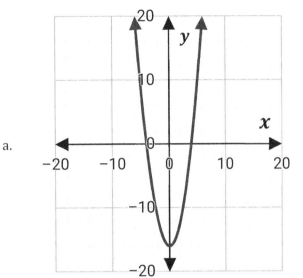

c.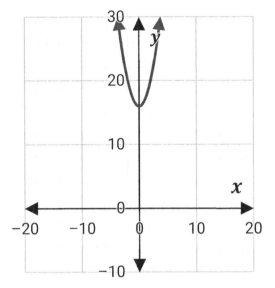

b.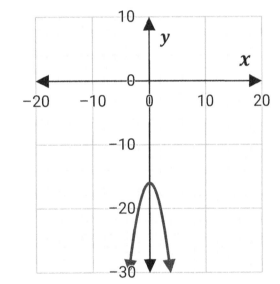

d.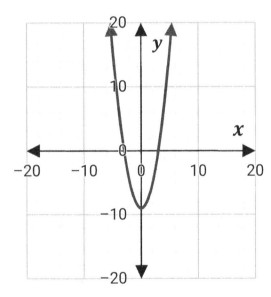

6. Which graph matches the equation $y = -x^2 - 2x - 1$?

a.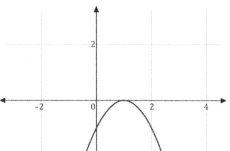

c.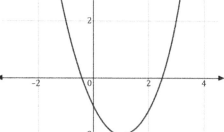

b.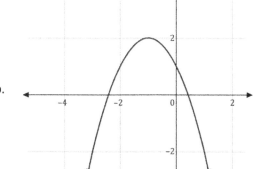

d.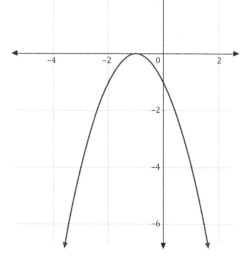

7. Which graph matches the equation $y = -x^2 + 6x + 5$?

a.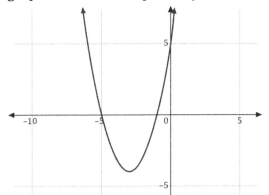

c.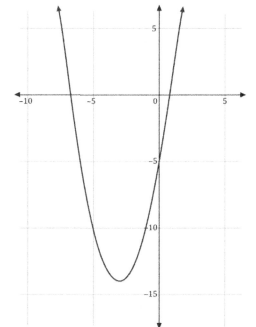

b.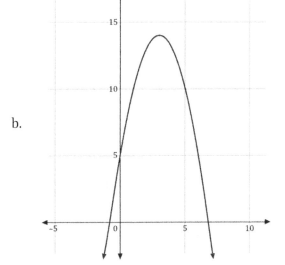

d.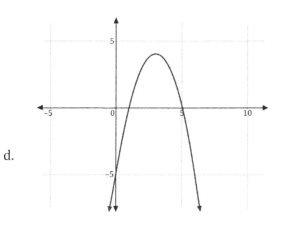

8. Which graph matches the equation $y = -2x^2 - 8x + 9$?

a.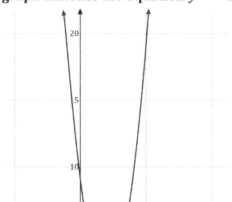

b.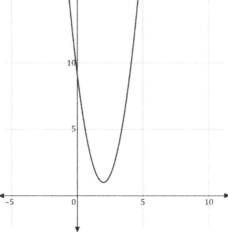

c.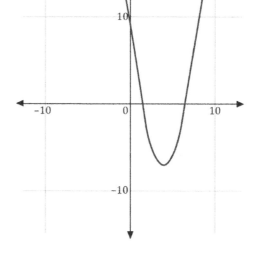

d.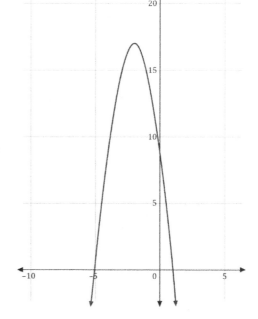

9. Which graph matches the equation $y = -\frac{1}{2}x^2 - 4x + ?$

a.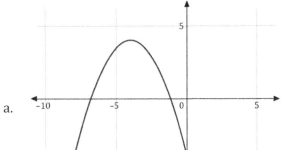

c.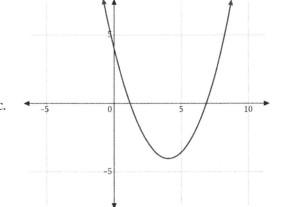

b.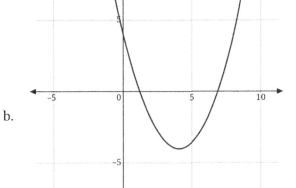

d.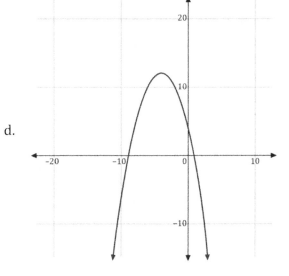

10. Which graph matches the equation $y = -2x^2 + 1$?

a.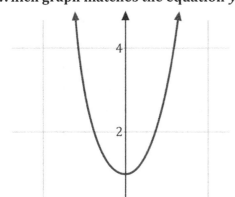

c.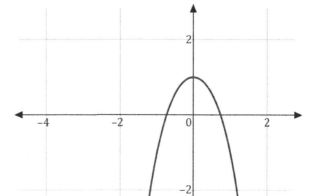

b.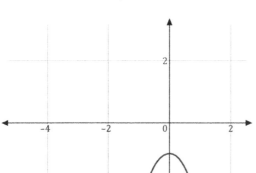

d.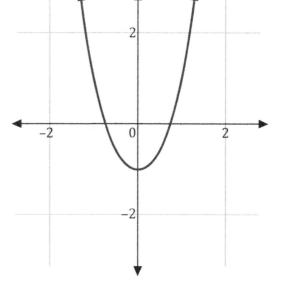

11. Which graph matches the equation $y = 3x^2 + 6x - 7$?

a.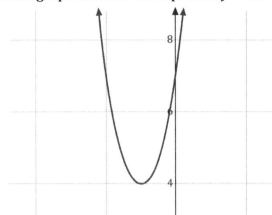

c.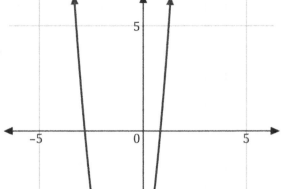

b.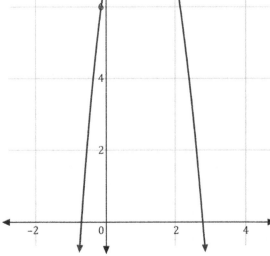

d.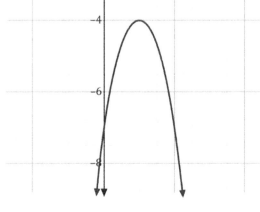

12. Which graph matches the equation $y = x^2 - 4x - 8$?

a.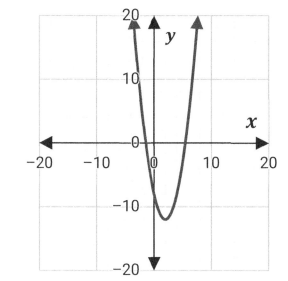

c.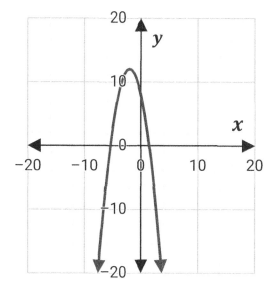

b.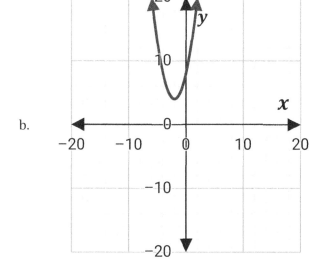

d.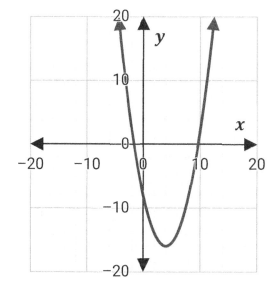

13. A physics class is experimenting with a model rocket. After s seconds, the rocket's height in feet, $h(s)$, is modeled by the equation $h(s) = -5s^2 + 20s + 3$. Graph the equation and identify the maximum height of the rocket.

a.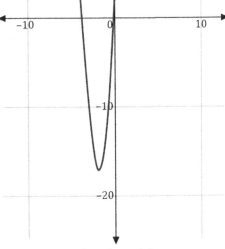
The maximum height is 3 feet.

b.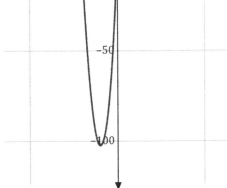
The maximum height is 23 feet.

c.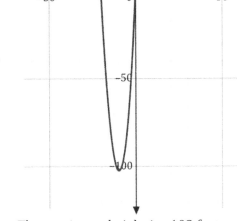
The maximum height is −103 feet.

d.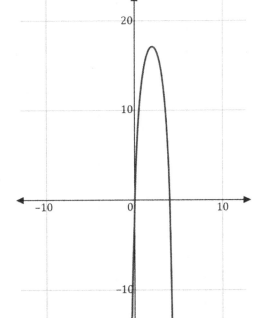
The maximum height is 17 feet.

14. Jack is at a carnival and wants to try the high striker, a game that tests your strength by determining how far in the air you can move a weight with a hammer. If the weight reaches 17 feet, you win the game. The equation $h(t) = -16t^2 + 32t + 1$ represents the height, $h(t)$, t seconds after Jack hits the weight. Graph the equation and find out whether Jack won the game.

a.
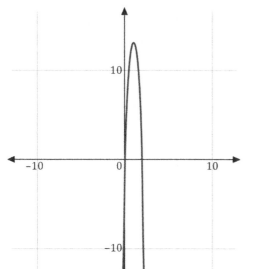
Jack's weight reached 13 feet, so he did not win the game.

c.
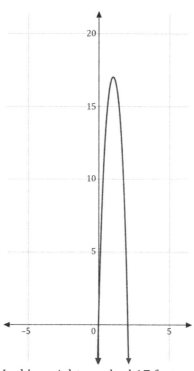
Jack's weight reached 17 feet, so he won the game.

b.
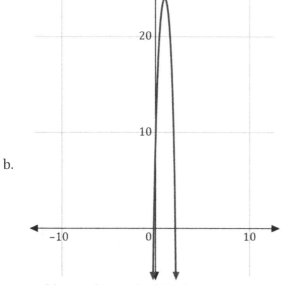
Jack's weight reached 24 feet, so he won the game.

d.
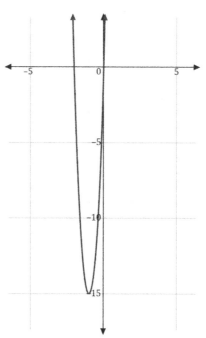
Jack's weight reached −15 feet, so he did not win the game.

15. A batter hits a baseball from ground level with an initial vertical velocity of 64 feet per second. The equation $y = -16x^2 + 64x$ represents the height of the baseball in the air with respect to the number of seconds that have passed. Based on this information, graph the equation to figure out the height of the baseball after it's been in the air for 3 seconds.

a.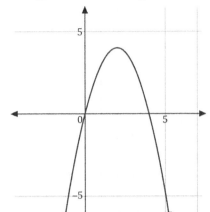
The height of the baseball at 3 seconds is 3 feet.

c.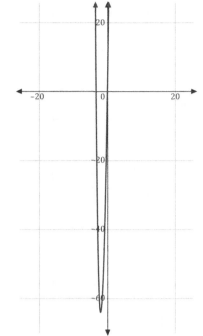
The height of the baseball at 3 seconds is −48 feet.

b.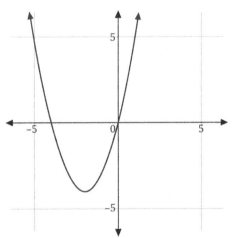
The height of the baseball at 3 seconds is −3 feet.

d.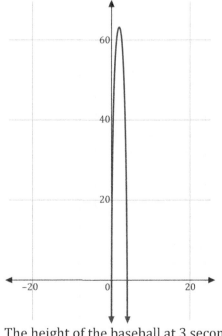
The height of the baseball at 3 seconds is 48 feet.

8.11 Manipulating Quadratic Functions

PRACTICE QUESTIONS

1. Which function represents the graph of $f(x) = x^2$ after it has been shifted 7 units down?
   a. $g(x) = -7x^2$
   b. $g(x) = x^2 - 7$
   c. $g(x) = (x - 7)^2$
   d. $g(x) = (-7x)^2$

2. Which function represents the graph of $f(x) = x^2$ after it is shifted 2 units to the right?
   a. $g(x) = x^2 + 2$
   b. $g(x) = x^2 - 2$
   c. $g(x) = (x + 2)^2$
   d. $g(x) = (x - 2)^2$

3. Which function represents the graph of $f(x) = x^2$ after it is stretched vertically by a factor of 6?
   a. $g(x) = x^2 + 6$
   b. $g(x) = 6x^2$
   c. $g(x) = (x + 6)^2$
   d. $g(x) = \frac{1}{6}x^2$

4. Which function represents the graph of $f(x) = x^2$ after it is compressed horizontally by a factor of $\frac{1}{2}$?
   a. $g(x) = (2x)^2$
   b. $g(x) = 2x^2$
   c. $g(x) = x^2 + 2$
   d. $g(x) = \left(\frac{x}{2}\right)^2$

5. Which function represents the graph of $f(x) = x^2$ after it is shifted 3 units down?
   a. $g(x) = x^2 - 3$
   b. $g(x) = -3x^2$
   c. $g(x) = (x - 3)^2$
   d. $g(x) = (x + 3)^2$

6. Which function represents the graph of $f(x) = x^2$ after it is shifted 5 units to the left?
   a. $g(x) = -5x^2$
   b. $g(x) = (x - 5)^2$
   c. $g(x) = (x + 5)^2$
   d. $g(x) = x^2 - 5$

7. Which function represents the graph of $f(x) = x^2$ after it is compressed vertically by a factor of $\frac{1}{2}$?

   a. $g(x) = x^2 + \frac{1}{2}$
   b. $g(x) = \frac{1}{2}x^2$
   c. $g(x) = \left(\frac{x}{2}\right)^2$
   d. $g(x) = \left(x + \frac{1}{2}\right)^2$

8. The function $f(x) = x^2$ is translated to create the function $g(x) = x^2 + 5$. Which statement describes the translation?

   a. The graph of $f(x)$ shifted up 5 units to create $g(x)$.
   b. The graph of $f(x)$ shifted down 5 units to create $g(x)$.
   c. The graph of $f(x)$ shifted left 5 units to create $g(x)$.
   d. The graph of $f(x)$ shifted right 5 units to create $g(x)$.

9. The function $f(x) = x^2$ is translated to create the function $g(x) = x^2 - 11$. Which statement describes the translation?

   a. The graph of $f(x)$ shifted up 11 units to create $g(x)$.
   b. The graph of $f(x)$ shifted down 11 units to create $g(x)$.
   c. The graph of $f(x)$ shifted left 11 units to create $g(x)$.
   d. The graph of $f(x)$ shifted right 11 units to create $g(x)$.

10. The function $f(x) = x^2$ is translated to create the function $g(x) = (x - 6)^2$. Which statement describes the translation?

    a. The graph of $f(x)$ shifted up 6 units to create $g(x)$.
    b. The graph of $f(x)$ shifted down 6 units to create $g(x)$.
    c. The graph of $f(x)$ shifted left 6 units to create $g(x)$.
    d. The graph of $f(x)$ shifted right 6 units to create $g(x)$.

11. The function $f(x) = x^2$ is translated to create the function $g(x) = (x + 9)^2$. Which statement describes the translation?

    a. The graph of $f(x)$ shifted up 9 units to create $g(x)$.
    b. The graph of $f(x)$ shifted down 9 units to create $g(x)$.
    c. The graph of $f(x)$ shifted left 9 units to create $g(x)$.
    d. The graph of $f(x)$ shifted right 9 units to create $g(x)$.

12. The graph of $f(x) = x^2$ is shown on Graph A. The graph of $f(x)$ is translated to create the graph of $g(x)$. What is the equation of Graph B?

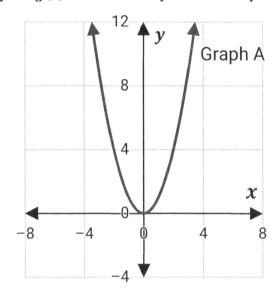
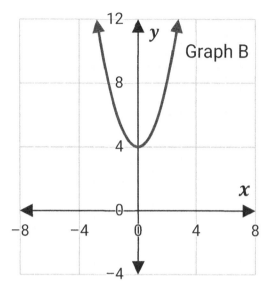

 a. $g(x) = x^2 + 4$
 b. $g(x) = x^2 - 4$
 c. $g(x) = (x + 4)^2$
 d. $g(x) = (x - 4)^2$

13. Jessica is a graphic designer. She needs to stretch an image vertically by a factor of 3. If the original image is represented by the function $f(x) = x^2$, which function represents this manipulation?

 a. $g(x) = (x - 3)^2$
 b. $g(x) = x^2 - 3$
 c. $g(x) = -3x^2$
 d. $g(x) = 3x^2$

14. An elementary school is ordering a new banner for their gymnasium. After seeing the design proof, they want to shift the logo's positioning. The original logo, represented by the function $f(x) = x^2$, needs to be moved 7 inches down. Which function represents this translation?

 a. $g(x) = -7x^2$
 b. $g(x) = (x - 7)^2$
 c. $g(x) = x - 7$
 d. $g(x) = x^2 - 7$

15. Jacob is a structural engineer. He is working on designing a new roller coaster for an amusement park. After creating an original design, he realizes he needs to change one of the roller coaster's dips. The dip, represented by the function $f(x) = x^2$, needs to be horizontally translated one foot to the right. Which function represents this translation?
   a. $g(x) = (x - 1)^2$
   b. $g(x) = (x + 1)^2$
   c. $g(x) = x^2 + 1$
   d. $g(x) = x^2 - 1$

Chapter 9: Exponential Functions

9.1 Domain and Range of Exponential Functions

PRACTICE QUESTIONS

1. Find the domain of the function $f(x) = 5 \cdot 1.3^x$.
   a. $x > 5$
   b. $x > 6.5$
   c. $-\infty < x < \infty$
   d. $x > 0$

2. Find the range of the function $f(x) = 7 \cdot 2^x$.
   a. $f(x) > 0$
   b. $f(x) \geq 7$
   c. $-\infty < f(x) < \infty$
   d. $f(x) > 2$

3. Find the domain of the function $s(t) = -2.7(0.4)^t$.
   a. $t < 0$
   b. $t \leq -2.7$
   c. $t > 0$
   d. $-\infty < t < \infty$

4. Find the range of the function $f(x) = -9.3 \cdot 5^x$.
   a. $-\infty < f(x) < \infty$
   b. $f(x) \leq -46.5$
   c. $f(x) \leq -9.3$
   d. $f(x) < 0$

5. Find the domain and range of the function $f(x) = 11 \cdot 10^x$.
   a. The domain is $-\infty < x < \infty$, and the range is $f(x) > 10$.
   b. The domain is $-\infty < x < \infty$, and the range is $f(x) > 0$.
   c. The domain is $-\infty < x < \infty$, and the range is $f(x) \geq 11$.
   d. The domain is $x \geq 0$, and the range is $f(x) \geq 10$.

6. Find the domain and range of the function $f(x) = -21 \cdot 7^x$.
   a. The domain is $-\infty < x < \infty$, and the range is $f(x) > 7$.
   b. The domain is $-\infty < x < \infty$, and the range is $f(x) \leq -21$.
   c. The domain is $-\infty < x < \infty$, and the range is $f(x) < 0$.
   d. The domain is $x \geq 0$, and the range is $f(x) \geq 7$.

7. Find the range of the function $f(x) = 2\left(\frac{1}{3}\right)^x + 5$.

    a. $0 < f(x) \leq 7$
    b. $-\infty < f(x) < \infty$
    c. $f(x) > 0$
    d. $f(x) > 5$

8. Find the domain of the function $f(x) = 11.3(0.4)^x - 12$.

    a. $x \geq 11.3$
    b. $-\infty < x < \infty$
    c. $x \geq -12$
    d. $x \geq 0$

9. Find the range of the function $f(x) = -8 \cdot 3^x + 14$.

    a. $f(x) > 14$
    b. $f(x) < 14$
    c. $f(x) < 0$
    d. $f(x) > -8$

10. Find the domain and range of the function $f(x) = -4\left(\frac{3}{8}\right)^x - 20$.

    a. The domain is $-\infty < x < \infty$, and the range is $f(x) > -20$.
    b. The domain is $x \leq 0$, and the range is $f(x) \leq -24$.
    c. The domain is $-\infty < x < \infty$, and the range is $f(x) \leq -26$.
    d. The domain is $-\infty < x < \infty$, and the range is $f(x) < -20$.

11. Find the range of the function $f(x) = 1.1 \cdot 4^x - 9$.

    a. $f(x) \geq -4.6$
    b. $f(x) > 0$
    c. $f(x) < -9$
    d. $f(x) > -9$

12. Find the range of the function $f(x) = -2.7 \cdot 3^x + 4$.

    a. $f(x) \leq -2.7$
    b. $f(x) > 3$
    c. $f(x) < 4$
    d. $f(x) \geq 4$

13. Today you invest $1,000 in an account that pays interest compounded continuously. You estimate the amount of money in the account t years after today using the function $A(t) = 1,000 \cdot 1.03^t$ dollars. What is the domain of this function?
   a. $t \geq 1.03$
   b. $t \geq 1,000$
   c. $t \geq 0$
   d. $-\infty < t < \infty$

14. Today you invest $2,500 in an account that pays interest compounded continuously. You estimate the amount of money in the account t years after today using the function $A(t) = 2,500 \cdot 1.02^t$ dollars. What is the range of this function?
   a. $A(t) \geq 2,500$
   b. $A(t) \geq 0$
   c. $-\infty < A(t) < \infty$
   d. $A(t) \geq 1.02$

15. At this moment you have a pure sample of 100 grams of a radioactive isotope. It undergoes radioactive decay at such a rate that in t days the amount of the isotope left is given by the function $A(t) = 100 \cdot 0.834^t$ grams. What is the range of this function?
   a. $0 < A(t) \leq 100$
   b. $A(t) > 0$
   c. $A(t) \leq 100$
   d. $A(t) \geq 100$

9.2 Writing Exponential Functions

Practice Questions

1. Jamie opens a bank account with $20. She doubles the amount of money in the bank account each month. This scenario is represented by the exponential function $f(x) = 20 \cdot 2^x$. Interpret the meaning of 20 in this function.
   a. 20 is the common ratio, which is the constant factor between consecutive terms in the sequence.
   b. 20 is the initial value, which is the original amount of money in Jamie's bank account.
   c. 20 is the common ratio, which is the original amount of money in Jamie's bank account.
   d. 20 is the initial value, which is the constant factor between consecutive terms in the sequence.

2. A baseball team sells 300 tickets to their most recent game. For each consecutive game they win, the number of tickets sold increases by a factor of 1.5. Which function represents the number of tickets bought after the baseball team has had x consecutive wins?
   a. $f(x) = 300 \cdot 1.5^x$
   b. $f(x) = 1.5 \cdot 300^x$
   c. $f(x) = x \cdot 300^{1.5}$
   d. $f(x) = 300 \cdot x^{1.5}$

3. Monica puts $3,000 in a savings account that earns 1% in interest each year. This scenario is represented by the exponential function $f(x) = 3,000 \cdot 1.01^x$. Interpret the meaning of 1.01 in this function.
   a. 1.01 is the amount of money earned in interest each year.
   b. 1.01 is the number of years the money is saved in the account.
   c. 1.01 is the initial value, which is the original amount of money in the savings account.
   d. 1.01 is the common ratio for the rate of exponential growth.

4. Melissa started a Spanish club at her high school. In its first year, the club had 15 members. Each year after, the number of club members increased by a factor of 1.1. Which function represents the number of members in the Spanish club x years after it started?
   a. $f(x) = 1.1 \cdot x^{15}$
   b. $f(x) = 15^x \cdot 0.1$
   c. $f(x) = 15 \cdot 1.1^x$
   d. $f(x) = 1.1 \cdot 15^x$

5. The population of Plainview can be modeled with the exponential function $f(x) = 30,000 \cdot 1.25^x$. Interpret the meaning of 30,000 in this function.
   a. 30,000 is the common ratio for the rate of exponential growth.
   b. 30,000 is the final value, which is the ending population of Plainview.
   c. 30,000 is the number of years during which the population of Plainview increases.
   d. 30,000 is the initial value, which is the original population of Plainview for which the model applies.

6. The table below shows the growth of a culture of bacteria. Which exponential function represents the number of bacteria x days after it started growing?

Bacteria	100	300	900	2,700	8,100
Day	0	1	2	3	4

   a. $f(x) = 100 \cdot 3^x$
   b. $f(x) = 3 \cdot 100^x$
   c. $f(x) = x \cdot 100^3$
   d. $f(x) = 100 \cdot x^3$

7. The world population in 2010 was approximately 6.92 billion. The annual rate of increase was about 1.2%. Assuming the rate of increase continues to be 1.2%, the exponential function that models the world population is $f(x) = 6.92 \cdot 1.012^x$. Interpret the meaning of x in this function.
   a. x stands for the initial value, which is the world population in 2010.
   b. x stands for the number of years the population increases after 2010.
   c. x stands for the number of people the population increases by.
   d. x stands for the common ratio for the rate of exponential growth.

8. The table of ordered pairs shows an exponential relationship. Which exponential function matches the table?

x	0	1	2	3	4
y	10	20	40	80	160

   a. $f(x) = 2 \cdot 10^x$
   b. $f(x) = 10 \cdot 10^x$
   c. $f(x) = 10 \cdot 2^x$
   d. $f(x) = 10 \cdot x^2$

9. A city with a population of 50,000 grows at a rate of 4% each year. Which function represents the population at the end of x years?
   a. $f(x) = 0.04 \cdot 50{,}000^x$
   b. $f(x) = 50{,}000 \cdot 0.04^x$
   c. $f(x) = 50{,}000 \cdot 1.04^x$
   d. $f(x) = 1.04 \cdot 50{,}000^x$

10. Margo invested $500 in a company's stock at the end of 2021. The value of the stock is predicted to increase at a rate of 10% each year. Which exponential function represents the growth in the value of the stock for x years?
    a. $f(x) = 1.10 \cdot 500^x$
    b. $f(x) = 10 \cdot 500^x$
    c. $f(x) = 500 \cdot 0.10^x$
    d. $f(x) = 500 \cdot 1.10^x$

11. John buys a car for $17,500. The value of the car depreciates at a rate of 9% per year. Which exponential function represents the value of the car after x years?

  a. $f(x) = 17{,}500 \cdot 0.91^x$
  b. $f(x) = 17{,}500 \cdot 1.09^x$
  c. $f(x) = 0.91 \cdot 17{,}500^x$
  d. $f(x) = 17{,}500 \cdot 9^x$

12. A new iPhone sells for $829. With each year that passes, the phone loses 20% of its original value. Which exponential function represents the phone's value x years after it is sold?

  a. $f(x) = 829 \cdot 0.2^x$
  b. $f(x) = 829 \cdot 80^x$
  c. $f(x) = 829 \cdot 20^x$
  d. $f(x) = 829 \cdot 0.8^x$

13. A painting is worth $500 and is expected to increase in value at a rate of 6% each year. Write an exponential function that represents this scenario. Then, find the expected value of the painting in 7 years. Round your answer to the nearest hundredth.

  a. $3,000.00
  b. $715.36
  c. $3,710.00
  d. $751.82

14. A laptop valued at $1,300 depreciates at a rate of 13% per year. Write an exponential function that represents this scenario. Then, find the expected value of the laptop after 2 years. Round your answer to the nearest hundredth.

  a. $9,700.00
  b. $983.97
  c. $1,287.00
  d. $659.97

15. An investment valued at $60,000 increases at a rate of 7.8% each year. Write an exponential function that represents this scenario. Then, find the value of the investment after 15 years. Round your answer to the nearest hundredth.

  a. $60,117.00
  b. $70,200.00
  c. $185,111.19
  d. $17,746.60

9.3 Graphing Exponential Functions and Determining Key Features

Practice Questions

1. Choose the graph of the exponential function $f(x) = 4 \cdot 2^x$.

a.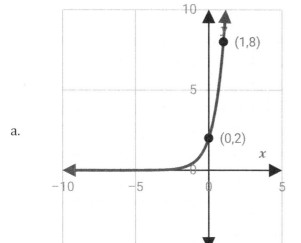

c.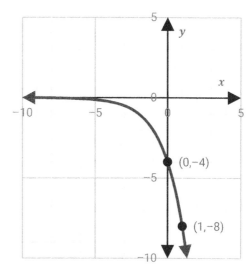

b.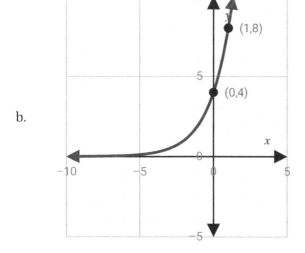

d.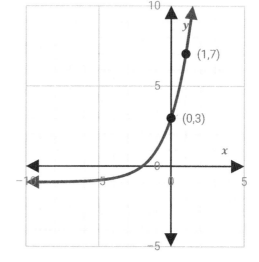

2. What is the y-intercept of the exponential function graphed on the coordinate plane below?

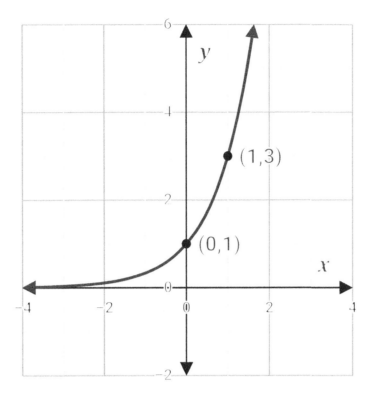

a. (1,3)
b. (0,1)
c. (0,0)
d. (−2,1)

3. Which of the following gives the general shape of the graph of the exponential function $f(x) = -4\left(\frac{2}{5}\right)^x$?

a.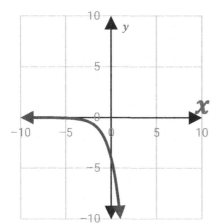

c.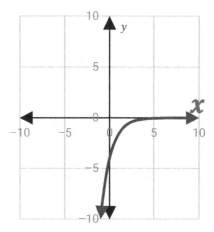

b.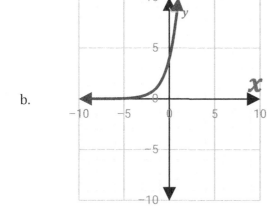

d.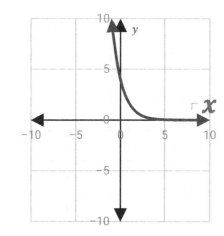

4. An exponential function is graphed on the coordinate plane shown below. What is the horizontal asymptote of the graph?

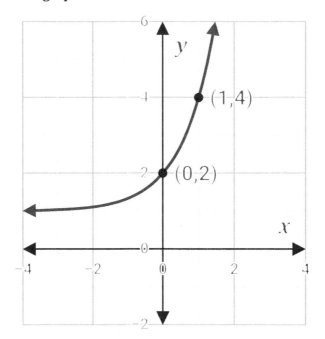

a. $y = 4$
b. $y = 2$
c. $y = 1$
d. $y = 0$

5. Choose the graph of the exponential function $f(x) = 6 \cdot \frac{1}{3}^x$.

a.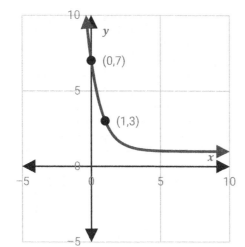

c.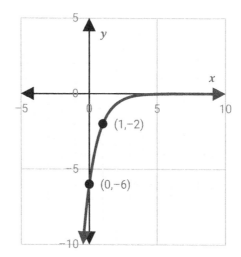

b.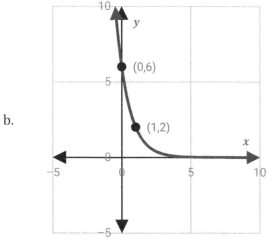

d.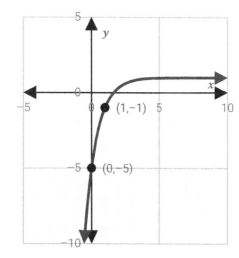

6. Which graph shows an exponential function with an initial value of $\frac{1}{2}$?

a.

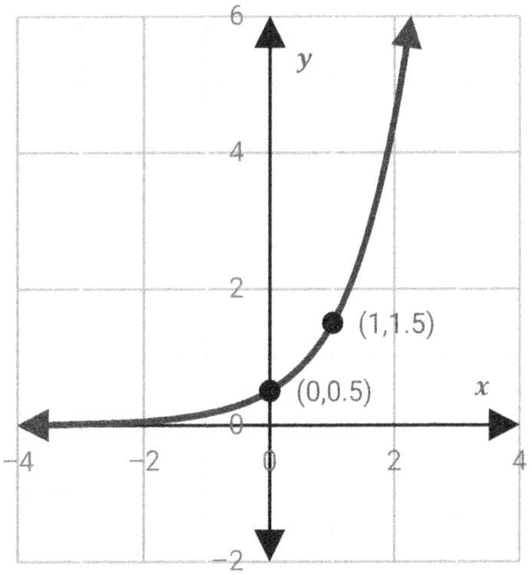

c.

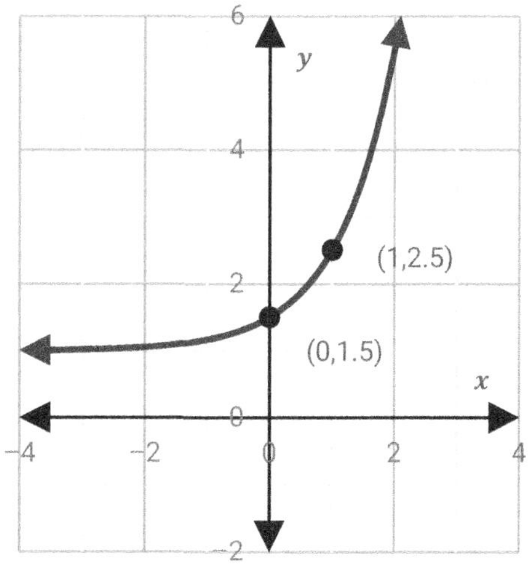

b.

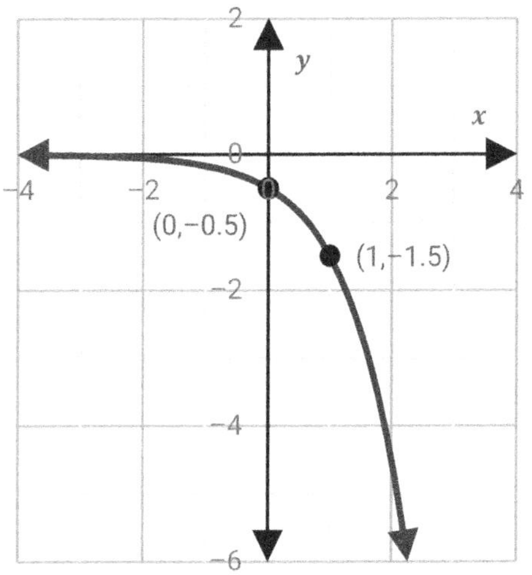

d.

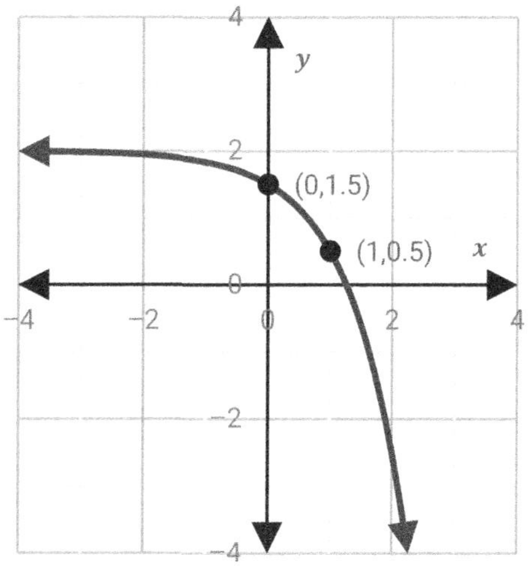

7. Based on the graph of the exponential function below, which statement is true?

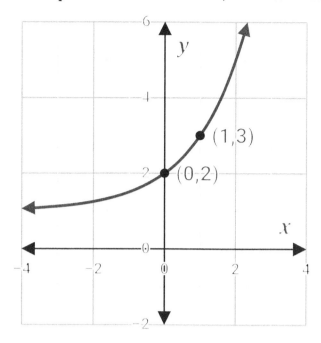

a. The graph is increasing, so the graph shows exponential growth.
b. The graph is increasing, so the graph shows exponential decay.
c. The graph is decreasing, so the graph shows exponential growth.
d. The graph is decreasing, so the graph shows exponential decay.

8. What is the horizontal asymptote of the graph of the exponential function shown?

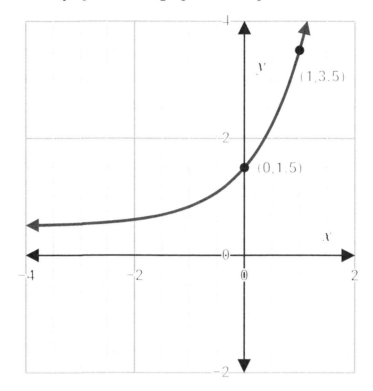

a. $y = 3.5$
b. $y = 1.5$
c. $y = -0.5$
d. $y = 0.5$

9. Choose the graph of the function $f(x) = \frac{2^x}{3}$.

a.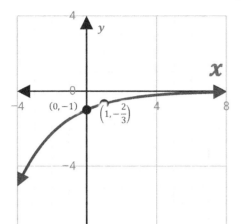

c.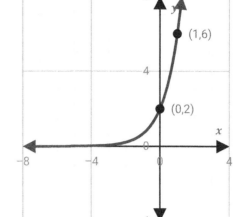

b.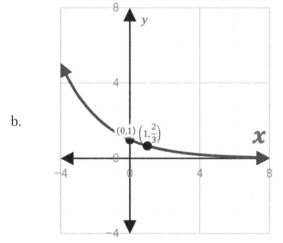

d.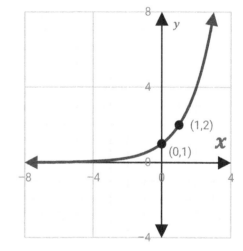

10. What is the y-intercept of the function shown?

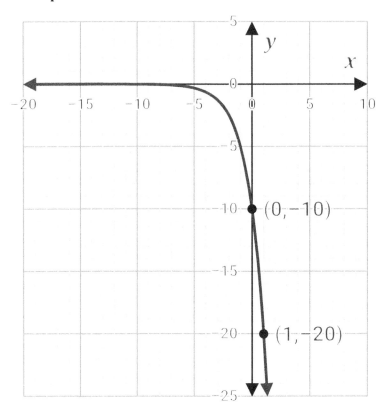

a. (0,0)
b. (1,−20)
c. (1,1)
d. (0,−10)

11. In the exponential function $f(x) = \frac{1}{4}^x + 2$, the base, b, of the exponent is $\frac{1}{4}$. What does this tell us about the function?

a. $b < 1$, so the function is a growth function.
b. $b < 1$, so the function is a decay function.
c. $0 < b < 1$, so the function is a growth function.
d. $0 < b < 1$, so the function is a decay function.

12. The graph of a function f is shown on the coordinate plane below. Which graph shows $f(x) + 3$?

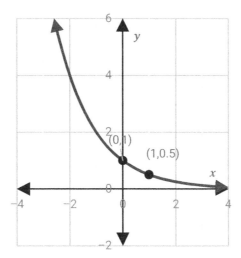

a.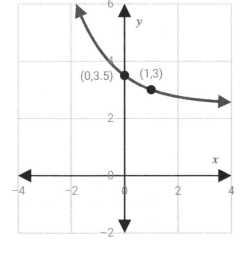

c.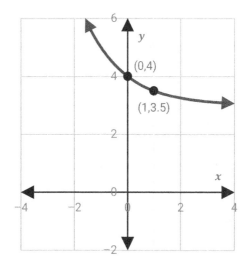

b.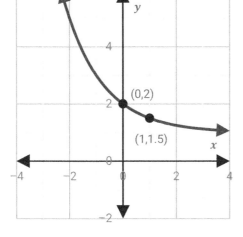

d.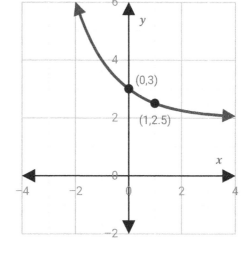

13. The rate at which a culture of bacteria self-replicates is modeled by the exponential function $f(x) = 9 \cdot 2^x$. In this function, x stands for time elapsed in minutes, and $f(x)$ stands for the number of bacteria cells. This function is graphed on the coordinate plane below. Based on this information, how many bacteria cells are present at 2 minutes?

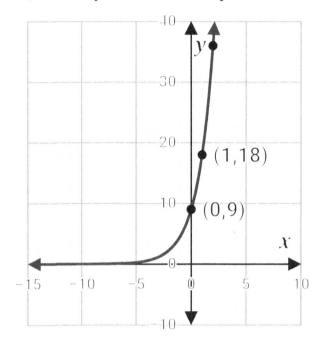

- a. 36 bacteria cells
- b. 18 bacteria cells
- c. 9 bacteria cells
- d. 1 bacteria cell

14. A population of 8 beavers tripled every month. This exponential relationship can be modeled by the function $f(x) = 8 \cdot 3^x$, where x stands for time elapsed in months, and $f(x)$ stands for the total number of beavers. Which graph matches this exponential function?

a.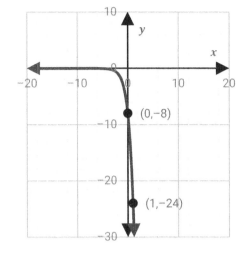

c.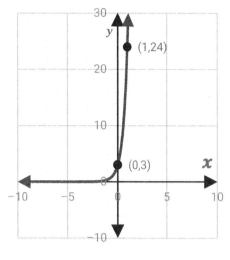

b.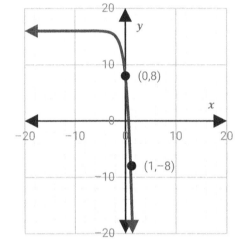

d.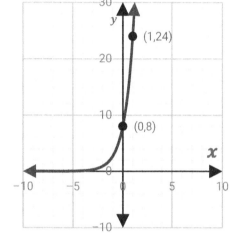

15. On his 18th birthday, Keith opens a bank account with $2,000. He doubles the amount of money in the account each year. This scenario can be modeled by the exponential function $f(x) = 2,000 \cdot 2^x$, where x stands for time elapsed in years, and $f(x)$ stands for Keith's total amount of money saved. Based on this information, how much money will Keith save after 3 years?

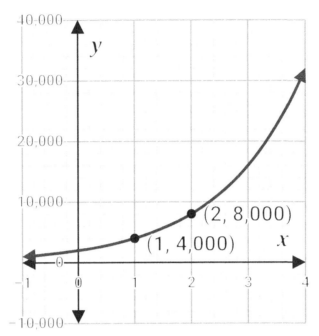

a. $4,000.00
b. $8,000.00
c. $16,000.00
d. $28,000.00

Chapter 10: Radical Expressions

10.1 Simplifying Radical Expressions

PRACTICE QUESTIONS

1. Simplify the radical expression $\sqrt{98}$.
   a. 49
   b. $7\sqrt{14}$
   c. $7\sqrt{2}$
   d. $49\sqrt{2}$

2. Simplify the radical expression $\sqrt{20}$.
   a. 10
   b. $4\sqrt{5}$
   c. $2\sqrt{10}$
   d. $2\sqrt{5}$

3. Simplify the radical expression $\sqrt{1,000}$.
   a. $2\sqrt{250}$
   b. $10\sqrt{10}$
   c. $100\sqrt{10}$
   d. $5\sqrt{40}$

4. Simplify the radical expression $\sqrt{128}$.
   a. $2\sqrt{32}$
   b. $4\sqrt{8}$
   c. $8\sqrt{2}$
   d. $2\sqrt{64}$

5. Simplify the radical expression $\sqrt{180}$.
   a. $2\sqrt{45}$
   b. $3\sqrt{20}$
   c. $6\sqrt{5}$
   d. $2\sqrt{90}$

6. Simplify the radical expression $\sqrt{162}$.
   a. $3\sqrt{18}$
   b. $9\sqrt{2}$
   c. $3\sqrt{2}$
   d. $81\sqrt{2}$

7. Simplify the radical expression $\sqrt{50x^3}$. Assume x is nonnegative.
   a. $25x$
   b. $5\sqrt{2x^2}$
   c. $x\sqrt{50}$
   d. $5x\sqrt{2x}$

8. Simplify the radical expression $\sqrt{64x}$. Assume x is nonnegative.
   a. $8\sqrt{x}$
   b. $2\sqrt{16x}$
   c. $4\sqrt{4x}$
   d. $8x$

9. Simplify the radical expression $\sqrt{216x^2}$. Assume x is nonnegative.
   a. $3x\sqrt{24}$
   b. $6x\sqrt{6}$
   c. $6\sqrt{6x^2}$
   d. $2x\sqrt{54}$

10. Simplify the radical expression $\sqrt{63x^2y^3}$. Assume x and y are nonnegative.
    a. $xy\sqrt{63y}$
    b. $3xy\sqrt{7y}$
    c. $3x\sqrt{7y^3}$
    d. $8xy\sqrt{y}$

11. Simplify the radical expression $\sqrt{30x^5y^7}$. Assume x and y are nonnegative.
    a. $x^2y^3\sqrt{30xy}$
    b. $2x^2y^3\sqrt{15xy}$
    c. $6xy\sqrt{5x^3y^5}$
    d. $xy\sqrt{30x^3y^5}$

12. Simplify the radical expression $\sqrt{121x^6y^3}$. Assume x and y are nonnegative.
    a. $11x^3y\sqrt{y}$
    b. $x^3y\sqrt{121y}$
    c. $11x\sqrt{x^3y^3}$
    d. $x^3\sqrt{121y^3}$

13. A square has an area of 300 square feet. Find the length of a side of the square. Simplify your answer.
    a. $10\sqrt{3}$ feet
    b. $50\sqrt{3}$ feet
    c. $4\sqrt{75}$ feet
    d. 75 feet

14. If a square has sides of length s, then a diagonal of the square has length $\sqrt{2s^2}$. Simplify the expression for the length of the diagonal.
    a. $2\sqrt{s}$
    b. $s\sqrt{2}$
    c. $2\sqrt{s^2}$
    d. $2s$

15. When a heavy, dense object is dropped and allowed to fall freely, the time that it takes to fall a distance of d feet is approximately $t = \sqrt{\frac{d}{16}}$ seconds. How long does it take such an object to fall 448 feet? Simplify your answer.
    a. $\sqrt{14}$ seconds
    b. $2\sqrt{14}$ seconds
    c. $8\sqrt{7}$ seconds
    d. $2\sqrt{7}$ seconds

10.2 Solving Equations Involving Radicals

Practice Questions

1. Solve the equation $\sqrt{3x-5} = 7$.
   a. $x = 4$
   b. $x = 16$
   c. $x = 18$
   d. $x = 21$

2. Solve the equation $\sqrt{2x-1} = 9$.
   a. $x = 5$
   b. $x = 18$
   c. $x = 24$
   d. $x = 41$

3. Solve the equation $4\sqrt{6x-2} + 5 = 13$.
   a. $x = 1$
   b. $x = 4$
   c. $x = 6$
   d. No real solutions

4. Solve the equation $2\sqrt{9x+4} - 4 = 10$.
   a. $x = 49$
   b. $x = 45$
   c. $x = 7$
   d. $x = 5$

5. Solve the equation $3\sqrt{2x+8} - 9 = 27$.
   a. $x = 68$
   b. $x = 144$
   c. $x = 12$
   d. $x = 136$

6. Solve the equation $\sqrt{4x+5} = \sqrt{x-10}$.
   a. $x = 5$
   b. $x = -15$
   c. $x = 15$
   d. No real solutions

7. Solve the equation $\sqrt{x-9}=4$.
    a. $x=25$
    b. $x=-5$
    c. $x=7$
    d. No real solutions

8. Solve the equation $\sqrt{x+17}=3$.
    a. $x=20$
    b. $x=-14$
    c. $x=26$
    d. $x=-8$

9. Solve the equation $x-\sqrt{8-2x}=0$.
    a. $x=-4$
    b. $x=2$ or $x=-4$
    c. $x=\frac{8}{3}$
    d. $x=2$

10. Solve the equation $\sqrt{2x-6}+3=x-3$.
    a. $x=3$ or $x=5$
    b. $x=5$
    c. $x=-3$ or $x=-5$
    d. $x=3$

11. Solve the equation $\sqrt{4x-13}=\sqrt{3x-1}$.
    a. $x=2$
    b. $x=-12$
    c. $x=-14$
    d. $x=12$

12. Solve the equation $\sqrt{2x^2+5x-7}-\sqrt{x^2+5x-3}=0$ for all real solutions.
    a. $x=-2$
    b. $x=2$
    c. $x=\sqrt{10}$
    d. $x=2$ or $x=-2$

13. Solve the equation: $\sqrt{4x-4}=6$.
    a. $x=40$
    b. $x=10$
    c. $x=36$
    d. No real solutions

14. Solve the equation: $2\sqrt{3x-8} + 4 = 12$.
   a. $x = 4$
   b. $x = 16$
   c. $x = 24$
   d. $x = 8$

15. You begin with a rectangular piece of paper that has an area of x square centimeters. By cutting off a piece whose area is 11 square centimeters, you produce a square in which each side measures 7 centimeters. Find x, the area of the original rectangle.
   a. $x = 60$ square inches
   b. $x = 18$ square inches
   c. $x = 38$ square inches
   d. $x = 2\sqrt{15}$ square inches

10.3 Adding and Subtracting Radical Expressions

Practice Questions

1. Simplify the expression $4\sqrt{6} + 8\sqrt{6}$.
   a. $12\sqrt{12}$
   b. 72
   c. $12\sqrt{6}$
   d. $24\sqrt{3}$

2. Simplify the expression $20\sqrt{42} + 30\sqrt{42} + 11\sqrt{42}$.
   a. $183\sqrt{14}$
   b. $103\sqrt{42}$
   c. $61\sqrt{126}$
   d. $61\sqrt{42}$

3. Simplify the expression $-12\sqrt{17x} + 9\sqrt{17x}$. Assume x is nonnegative.
   a. $-51x$
   b. $-\sqrt{34x}$
   c. $-72\sqrt{17x}$
   d. $-3\sqrt{17x}$

4. Simplify the expression $10\sqrt{10} - 5\sqrt{10}$.
   a. $5\sqrt{10}$
   b. $-50\sqrt{10}$
   c. $10\sqrt{5}$
   d. 5

5. Simplify the expression $12\sqrt{21} - 16\sqrt{21} + \sqrt{21}$.
   a. $-4\sqrt{21}$
   b. $-3\sqrt{21}$
   c. $-5\sqrt{21}$
   d. $-4 + \sqrt{21}$

6. Simplify the expression $(8\sqrt{3} - 10\sqrt{5y}) - (7\sqrt{3} - 12\sqrt{5y})$. Assume y is nonnegative.
   a. 3
   b. $\sqrt{3} + 2\sqrt{5y}$
   c. $\sqrt{6} + 2\sqrt{10y}$
   d. $3\sqrt{3 + 5y}$

7. Simplify the expression $14\sqrt{2} + 3\sqrt{50}$.
   a. $34\sqrt{13}$
   b. $89\sqrt{2}$
   c. $29\sqrt{2}$
   d. 58

8. Simplify the expression $-15\sqrt{40} + 25\sqrt{10}$.
   a. $10\sqrt{30}$
   b. $-5\sqrt{10}$
   c. $-35\sqrt{10}$
   d. $50\sqrt{2}$

9. Simplify the expression $9y\sqrt{3y} + 4\sqrt{27y^3}$. Assume y is nonnegative.
   a. $13\sqrt{30(y+y^3)}$
   b. $45y\sqrt{3y}$
   c. $21y\sqrt{3y}$
   d. $117y^2$

10. Simplify the expression $18\sqrt{7} - 8\sqrt{63}$.
    a. $-6\sqrt{7}$
    b. -6
    c. $20\sqrt{14}$
    d. $-54\sqrt{7}$

11. Simplify the expression $16\sqrt{132} - 31\sqrt{33}$.
    a. $\sqrt{33}$
    b. 1
    c. $-45\sqrt{11}$
    d. $33\sqrt{33}$

12. Simplify the expression $26x^2\sqrt{5x} - 2\sqrt{500x^5}$. Assume x is nonnegative.
    a. $6x^2\sqrt{10x}$
    b. $6x^2\sqrt{5x}$
    c. $6x^2$
    d. $-174\sqrt{5x}$

13. A rectangle has length $\sqrt{10}$ cm and width $\sqrt{90}$ cm. Find the perimeter of the rectangle. Simplify your answer.
    a. $4\sqrt{10}$ cm
    b. $8\sqrt{10}$ cm
    c. $20\sqrt{10}$ cm
    d. 40 cm

14. You are constructing two small square pens, each with an area of 35 square feet, and one large square pen with an area of 140 square feet. How many feet of fencing will you need altogether? Give an exact answer.
    a. $24\sqrt{35}$ ft
    b. $4\sqrt{35}$ ft
    c. $16\sqrt{35}$ ft
    d. $16\sqrt{210}$ ft

15. When a heavy, dense object is dropped and allowed to fall freely, the time that it takes to fall a distance of d feet is approximately $t = \sqrt{\frac{d}{16}}$ seconds. You drop a brick from the top of a tower. How much time passes between the moment when the brick is 32 feet below you and the moment when it is 128 feet below you? Give an exact answer.
    a. $\sqrt{2}$ seconds
    b. $\sqrt{6}$ seconds
    c. 2 seconds
    d. $\frac{\sqrt{2}}{4}$ seconds

10.4 Multiplying Radical Expressions

PRACTICE QUESTIONS

1. Simplify the product $3\sqrt{11} \cdot 4\sqrt{14}$.
    a. $24\sqrt{38}$
    b. $12\sqrt{154}$
    c. $24\sqrt{39}$
    d. 60

2. Simplify the product $5\sqrt{6} \cdot 8\sqrt{14}$.
    a. $80\sqrt{21}$
    b. $80\sqrt{5}$
    c. $160\sqrt{21}$
    d. $40\sqrt{84}$

3. Simplify the product $(-4\sqrt{30}) \cdot (2\sqrt{6})$.
    a. -48
    b. $-48\sqrt{5}$
    c. $-288\sqrt{5}$
    d. $-16\sqrt{6}$

4. Simplify the product $3\sqrt{2} \cdot \sqrt{6}$.
    a. $6\sqrt{3}$
    b. $12\sqrt{3}$
    c. $8\sqrt{3}$
    d. $6\sqrt{2}$

5. Simplify the product $3\sqrt{10} \cdot 2\sqrt{15}$.
    a. $150\sqrt{6}$
    b. 30
    c. $30\sqrt{6}$
    d. $25\sqrt{6}$

6. Simplify the product $(5\sqrt{7})(3\sqrt{70})$.
    a. $15\sqrt{77}$
    b. $515\sqrt{10}$
    c. $56\sqrt{10}$
    d. $105\sqrt{10}$

7. Simplify the product $(10\sqrt{3}) \cdot (-2\sqrt{21})$.
   a. $-180\sqrt{7}$
   b. $-60\sqrt{7}$
   c. $-40\sqrt{6}$
   d. $-60\sqrt{2}$

8. Simplify the product $\sqrt{2}(\sqrt{14} + \sqrt{22})$.
   a. $2\sqrt{7} + 2\sqrt{11}$
   b. $4 + 2\sqrt{6}$
   c. $6\sqrt{2}$
   d. $4\sqrt{7} + 4\sqrt{11}$

9. Simplify the product $\sqrt{6}(\sqrt{10} + \sqrt{15})$.
   a. $2\sqrt{15} + 3\sqrt{10}$
   b. $4\sqrt{15} + 9\sqrt{10}$
   c. $\sqrt{60} + \sqrt{90}$
   d. $5\sqrt{6}$

10. Simplify the product $(2 + \sqrt{10})(3 + \sqrt{10})$.
    a. $6 + 7\sqrt{10}$
    b. $16 + 5\sqrt{10}$
    c. 16
    d. $6 + 15\sqrt{10}$

11. Simplify the product $(6 - \sqrt{15})(5 - \sqrt{15})$.
    a. $30 - 9\sqrt{15}$
    b. $15 - 11\sqrt{15}$
    c. $30 - 13\sqrt{15}$
    d. $45 - 11\sqrt{15}$

12. Simplify the product $(7 + \sqrt{2})(7 - \sqrt{2})$.
    a. $49 + 14\sqrt{2}$
    b. 45
    c. 47
    d. $49 - 14\sqrt{2}$

13. A rectangular circuit board has a length of $4\sqrt{35}$ mm and a width of $2\sqrt{5}$ mm. Find the exact area of the circuit board. Simplify your answer.

- a. $40\sqrt{7}$ mm^2
- b. $30\sqrt{7}$ mm^2
- c. $200\sqrt{7}$ mm^2
- d. $16\sqrt{10}$ mm^2

14. The height of a triangular window from the base to the top of the window is $\sqrt{33}$ feet and the base of the window is $6\sqrt{3}$ feet long. Find the exact area of the window. Simplify your answer.

- a. 18 square feet
- b. $9\sqrt{11}$ square feet
- c. $3\sqrt{99}$ square feet
- d. $27\sqrt{11}$ square feet

15. A rectangular sheet of origami paper has a length of $8 + \sqrt{3}$ cm and a width of $6 + \sqrt{3}$ cm. Find the exact area of the sheet. Simplify your answer.

- a. $51 + 16\sqrt{3}$ cm^2
- b. $51 + 14\sqrt{3}$ cm^2
- c. 51 cm^2
- d. $14 + 2\sqrt{3}$ cm^2

10.5 Dividing Radical Expressions

PRACTICE QUESTIONS

1. Simplify the quotient $\frac{\sqrt{20}}{10}$.

   a. $\frac{2\sqrt{5}}{5}$
   b. $\frac{\sqrt{10}}{5}$
   c. $\sqrt{2}$
   d. $\frac{\sqrt{5}}{5}$

2. Simplify the quotient $\frac{5\sqrt{72}}{9}$.

   a. $\frac{10\sqrt{2}}{3}$
   b. $20\sqrt{2}$
   c. $\frac{5\sqrt{8}}{3}$
   d. $10\sqrt{2}$

3. Simplify the quotient $\frac{\sqrt{50x^3}}{15x}$. Assume x is positive.

   a. $\frac{x^2\sqrt{10}}{3}$
   b. $\frac{x\sqrt{10}}{3}$
   c. $\frac{\sqrt{2x}}{3}$
   d. $\frac{5\sqrt{2x}}{3}$

4. Simplify the quotient $\frac{14}{\sqrt{28}}$.

   a. $\frac{7}{\sqrt{7}}$
   b. $\frac{14}{\sqrt{7}}$
   c. $\frac{\sqrt{7}}{7}$
   d. $7\sqrt{7}$

5. Simplify the quotient $\frac{18}{\sqrt{36}}$. Assume x is positive.

   a. $\frac{1}{3}$
   b. 18
   c. 3
   d. $\frac{1}{2}$

6. Simplify the quotient $\frac{3x}{\sqrt{6x^2}}$. Assume x is positive.

   a. $\frac{\sqrt{6}}{2}$

   b. $\frac{3x}{\sqrt{6}}$

   c. $\frac{\sqrt{2}}{2}$

   d. $\frac{3}{\sqrt{6}}$

7. Simplify the quotient $\frac{\sqrt{8}+\sqrt{18}}{\sqrt{32}}$.

   a. $2\sqrt{2}$

   b. $\frac{\sqrt{13}}{4}$

   c. $\frac{5}{4}$

   d. $\frac{3\sqrt{2}}{2}$

8. Simplify the quotient $\frac{\sqrt{15}}{\sqrt{21}}$.

   a. $\frac{5}{\sqrt{7}}$

   b. $\frac{\sqrt{5}}{7}$

   c. $\frac{\sqrt{315}}{21}$

   d. $\frac{\sqrt{5}}{\sqrt{7}}$

9. Simplify the quotient $\frac{2\sqrt{3}}{3\sqrt{6}}$.

   a. $\frac{2}{\sqrt{6}}$

   b. $\frac{2}{6}$

   c. $\frac{2}{3\sqrt{3}}$

   d. $\frac{2}{3\sqrt{2}}$

10. Simplify the quotient $\frac{3\sqrt{105}}{8\sqrt{30}}$.

   a. $\frac{21}{16}$

   b. $\frac{3\sqrt{14}}{8}$

   c. $\frac{3\sqrt{7}}{8\sqrt{2}}$

   d. $\frac{\sqrt{42}}{16}$

11. Simplify the quotient $\frac{8\sqrt{12}-5\sqrt{27}}{4\sqrt{6}}$.
    a. $2\sqrt{2} - 15\sqrt{3}$
    b. $\frac{1}{4\sqrt{2}}$
    c. $\frac{3\sqrt{15}}{4\sqrt{2}}$
    d. $-\frac{15\sqrt{2}}{8}$

12. Simplify the quotient $\frac{9\sqrt{12x^3}}{2\sqrt{15x}}$. Assume x is positive.
    a. $\frac{9x^2}{\sqrt{5}}$
    b. $\frac{18x}{\sqrt{5}}$
    c. $\frac{18x^2}{\sqrt{5}}$
    d. $\frac{9x}{\sqrt{5}}$

13. The length of one side of the triangle is $\frac{8\sqrt{2}}{\sqrt{16}}$ inches. Simplify this measurement by rationalizing the denominator.
    a. $\frac{\sqrt{2}}{2}$ inches
    b. $\frac{1}{2}$ inch
    c. $2\sqrt{2}$ inches
    d. $4\sqrt{2}$ inches

14. Jessica took a math quiz on rationalizing denominators. When she got her graded quiz paper returned, she saw that her teacher wrote her score with a radical in the denominator. Jessica scored $\frac{\sqrt{36}}{\sqrt{49}}$ on her quiz. Simplify this expression by rationalizing the denominator to figure out Jessica's quiz score.
    a. Jessica got 6 out of 7 questions correct on her quiz.
    b. Jessica got 36 out of 49 questions correct on her quiz.
    c. Jessica got 49 out of 36 questions correct on her quiz.
    d. Jessica got 7 out of 6 questions correct on her quiz.

15. The two diagonals of a parallelogram have lengths $\sqrt{22}$ cm and $\sqrt{26}$ cm. Find the ratio of the longer diagonal to the shorter diagonal.
    a. $\frac{\sqrt{13}}{\sqrt{11}}$
    b. $\frac{\sqrt{11}}{\sqrt{13}}$
    c. $\sqrt{3}$
    d. $\frac{2\sqrt{6}}{11}$

10.6 Rationalizing the Denominator

PRACTICE QUESTIONS

1. Rationalize the denominator. If possible, simplify the fraction.

$$\frac{7}{\sqrt{2}}$$

a. $\frac{7}{4}$
b. $\frac{\sqrt{2}}{7}$
c. $\frac{7\sqrt{2}}{4}$
d. $\frac{7\sqrt{2}}{2}$

2. Rationalize the denominator. If possible, simplify the fraction.

$$\frac{2}{\sqrt{7}}$$

a. $\frac{2\sqrt{7}}{14}$
b. $\frac{2\sqrt{7}}{7}$
c. $\frac{2\sqrt{7}}{49}$
d. $\frac{14}{49}$

3. Rationalize the denominator. If possible, simplify the fraction.

$$\frac{12}{\sqrt{3}}$$

a. $2\sqrt{3}$
b. $4\sqrt{3}$
c. $\frac{4\sqrt{3}}{3}$
d. 12

4. Rationalize the denominator. If possible, simplify the fraction.

$$\frac{25}{\sqrt{5}}$$

a. $5\sqrt{5}$
b. $\sqrt{5}$
c. $\frac{25\sqrt{5}}{10}$
d. $2.5\sqrt{5}$

5. Rationalize the denominator. If possible, simplify the fraction.

$$\frac{12}{\sqrt{10}}$$

a. $\frac{12\sqrt{5}}{5}$
b. $\frac{12\sqrt{10}}{5}$
c. $\frac{6\sqrt{10}}{5}$
d. $\frac{6\sqrt{5}}{5}$

6. Rationalize the denominator. If possible, simplify the fraction.

$$\frac{1}{3\sqrt{2}}$$

a. $\frac{\sqrt{2}}{6}$
b. $\frac{\sqrt{1}}{3}$
c. $6\sqrt{2}$
d. $2\sqrt{6}$

7. Rationalize the denominator. If possible, simplify the fraction.

$$\frac{9}{\sqrt{15}}$$

a. $\frac{3\sqrt{15}}{5}$
b. $\frac{9\sqrt{15}}{30}$
c. $\frac{9\sqrt{15}}{17}$
d. $9\sqrt{15}$

8. Rationalize the denominator. If possible, simplify the fraction.

$$\frac{5}{2\sqrt{3}}$$

a. $\frac{5\sqrt{3}}{12}$
b. $\frac{5\sqrt{3}}{6}$
c. $\frac{5}{2}$
d. $\frac{5}{4}$

9. Rationalize the denominator. If possible, simplify the fraction.

$$\frac{4}{\sqrt{6}}$$

a. $\frac{\sqrt{6}}{2}$

b. $\frac{\sqrt{6}}{9}$

c. $\frac{\sqrt{6}}{3}$

d. $\frac{2\sqrt{6}}{3}$

10. Rationalize the denominator. If possible, simplify the fraction.

$$\frac{\sqrt{7}}{\sqrt{5}}$$

a. $\frac{\sqrt{35}}{5}$

b. $\frac{\sqrt{7}}{5}$

c. $\frac{\sqrt{7}}{25}$

d. $\frac{\sqrt{35}}{10}$

11. Rationalize the denominator. If possible, simplify the fraction.

$$\frac{5}{\sqrt{10}}$$

a. $\frac{\sqrt{10}}{20}$

b. $\frac{5\sqrt{10}}{12}$

c. $\frac{\sqrt{10}}{2}$

d. $\frac{\sqrt{10}}{4}$

12. Rationalize the denominator. If possible, simplify the fraction.

$$\frac{7}{\sqrt{12}}$$

a. $\frac{7\sqrt{3}}{12}$

b. $\frac{7\sqrt{12}}{24}$

c. $\frac{7\sqrt{3}}{6}$

d. $\frac{7\sqrt{12}}{144}$

13. A rectangular printed circuit board has an area of 100 mm² and a width of √70 mm. Find the exact length of the circuit board.

   a. $10\sqrt{10}$ mm
   b. $\frac{10\sqrt{70}}{7}$ mm
   c. $100\sqrt{70}$ mm
   d. $\frac{10\sqrt{7}}{7}$ mm

14. Christopher's family is eating pizza for dinner. While they're eating, Christopher gives them a math riddle. He tells them that he estimates that they have already eaten $\frac{8}{3\sqrt{10}}$ of the pizza. Simplify this expression by rationalizing the denominator.

   a. $\frac{4\sqrt{10}}{30}$
   b. $\frac{4\sqrt{10}}{15}$
   c. $\frac{8\sqrt{10}}{30}$
   d. $\frac{8\sqrt{10}}{15}$

15. A rock dropped from the top of a cliff falls $16t$ feet in $\sqrt{t}$ seconds ($t > 0$). The average speed of the rock equals the total distance fallen divided by the total time required for the fall. Find the average speed of the rock during this time.

   a. $4t$ feet per second
   b. $16\sqrt{t}$ feet per second
   c. $16t^2$ feet per second
   d. $4\sqrt{t}$ feet per second

10.7 Simplifying Radical Expressions with Binomial Denominators

PRACTICE QUESTIONS

1. Simplify the expression by rationalizing the denominator.

$$\frac{1}{\sqrt{2}+1}$$

   a. $\sqrt{2}+1$
   b. $\sqrt{2}-1$
   c. $1-\sqrt{2}$
   d. $\sqrt{2}$

2. Simplify the expression by rationalizing the denominator.

$$\frac{4}{3+\sqrt{5}}$$

   a. $\sqrt{3}-5$
   b. $\sqrt{5}-3$
   c. $3-\sqrt{5}$
   d. $3+\sqrt{5}$

3. Simplify the expression by rationalizing the denominator.

$$\frac{2}{\sqrt{6}-3}$$

   a. $-\frac{2\sqrt{6}+3}{3}$
   b. $\frac{2\sqrt{6}+3}{3}$
   c. $-\frac{2\sqrt{6}+6}{3}$
   d. $\frac{2\sqrt{6}+6}{3}$

4. Simplify the expression by rationalizing the denominator.

$$\frac{4}{\sqrt{2}-2}$$

   a. $2\sqrt{2}-4$
   b. $-2\sqrt{2}-4$
   c. $-2\sqrt{2}+4$
   d. $2\sqrt{2}+4$

5. Simplify the expression by rationalizing the denominator.
$$\frac{3}{1+\sqrt{3}}$$

a. $\frac{3\sqrt{3}}{2}$
b. $\frac{3-3\sqrt{3}}{2}$
c. $-\frac{3\sqrt{3}}{2}$
d. $-\frac{3-3\sqrt{3}}{2}$

6. Simplify the expression by rationalizing the denominator.
$$\frac{3}{2-\sqrt{5}}$$

a. $-6-3\sqrt{5}$
b. $6+3\sqrt{5}$
c. $3\sqrt{5}-6$
d. $\frac{6+3\sqrt{5}}{-21}$

7. Simplify the expression by rationalizing the denominator.
$$\frac{5}{\sqrt{2}-\sqrt{3}}$$

a. $-5\sqrt{2}-5\sqrt{3}$
b. $5\sqrt{2}+5\sqrt{3}$
c. $5\sqrt{2}-5\sqrt{3}$
d. $-5\sqrt{2}+5\sqrt{3}$

8. Simplify the expression by rationalizing the denominator.
$$\frac{\sqrt{6}}{\sqrt{2}+3}$$

a. $-\frac{2\sqrt{3}-3\sqrt{6}}{7}$
b. $\frac{2\sqrt{3}-3\sqrt{6}}{7}$
c. $-\frac{3\sqrt{6}-2\sqrt{3}}{5}$
d. $\frac{2\sqrt{3}-3\sqrt{6}}{5}$

9. Simplify the expression by rationalizing the denominator.
$$\frac{2}{\sqrt{5} + 2\sqrt{3}}$$

   a. $\frac{2\sqrt{5}+4\sqrt{3}}{7}$
   b. $-\frac{2\sqrt{5}+4\sqrt{3}}{7}$
   c. $\frac{2\sqrt{5}-4\sqrt{3}}{7}$
   d. $-\frac{2\sqrt{5}-4\sqrt{3}}{7}$

10. Simplify the expression by rationalizing the denominator.
$$\frac{\sqrt{3}}{5 + 2\sqrt{3}}$$

   a. $\frac{5\sqrt{3}-6}{37}$
   b. $\frac{5\sqrt{3}+6}{13}$
   c. $\frac{5\sqrt{3}-6}{13}$
   d. $\frac{5\sqrt{3}-23}{13}$

11. Simplify the expression by rationalizing the denominator.
$$\frac{-1 + \sqrt{3}}{\sqrt{3} - 1}$$

   a. 2
   b. 1
   c. $\frac{-\sqrt{3}}{2}$
   d. $\frac{\sqrt{3}}{2}$

12. Simplify the expression by rationalizing the denominator.
$$\frac{\sqrt{5} + 3}{4 - \sqrt{5}}$$

   a. $\frac{12\sqrt{5}+17}{11}$
   b. $\frac{7\sqrt{5}+17}{11}$
   c. $\frac{12\sqrt{5}+60}{11}$
   d. $\frac{7\sqrt{5}+60}{11}$

13. The margin of error is a statistic that tells the amount of random sampling error in a survey's results. A poll before a local election predicted that the incumbent candidate would win the election, with a margin of error of $\frac{2-\sqrt{3}}{4+\sqrt{3}}$. Rationalize the binomial denominator.

    a. $\frac{11-6\sqrt{3}}{13}$

    b. $\frac{11-6\sqrt{3}}{19}$

    c. $\frac{11+2\sqrt{3}}{13}$

    d. $\frac{11+2\sqrt{3}}{19}$

14. Holly and Theo are working on rationalizing binomial denominators. They each simplify the expression $\frac{\sqrt{3}-2}{\sqrt{3}+1}$, but they have come up with different answers. Holly says that after rationalizing the denominator and simplifying, there are 2 radicals in the numerator. Theo says that his answer only has 1 radical in the numerator. What is the simplified form of this expression, and who is correct?

    a. Holly is correct. The simplified form of the expression is $\frac{3-\sqrt{3}-2\sqrt{3}+2}{2}$.

    b. Theo is correct. The simplified form of the expression is $\frac{3-\sqrt{3}-2\sqrt{3}+2}{2}$.

    c. Holly is correct. The simplified form of the expression is $\frac{5-3\sqrt{3}}{2}$.

    d. Theo is correct. The simplified form of the expression is $\frac{5-3\sqrt{3}}{2}$.

15. The length of a rectangle is $\frac{5\sqrt{2}}{2\sqrt{5}-3}$ centimeters. Simplify this measurement by rationalizing the denominator.

    a. $\frac{25\sqrt{12}}{36}$ centimeters

    b. $\frac{25\sqrt{12}}{11}$ centimeters

    c. $\frac{10\sqrt{10}+15\sqrt{2}}{11}$ centimeters

    d. $\frac{10\sqrt{10}+15\sqrt{2}}{36}$ centimeters

Practice Test #1

1. Solve the inequality and graph your answer: $-2(3x - 5) \geq 58$.

   a. $x < -8$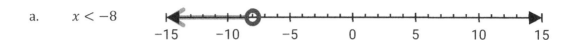

   b. $x \leq -8$

   c. $x \geq -8$

   d. $x > -8$

2. What is the value of x for the linear function $y = 3x - 7$ when the value of y is 5?
   a. -2
   b. 1
   c. 4
   d. 8

3. Rationalize the denominator in the fraction $\frac{30}{\sqrt{42}}$. Simplify your answer.

   a. $\frac{5\sqrt{42}}{7}$
   b. $\frac{5\sqrt{21}}{7}$
   c. $5\sqrt{6}$
   d. $\frac{10\sqrt{21}}{7}$

4. Simplify $\frac{9xy^5}{12x^4y^3}$.

   a. $\frac{3y^8}{4x^5}$
   b. $\frac{3y^2}{4x^4}$
   c. $\frac{3y^2}{4x^3}$
   d. $\frac{3x^3}{4y^2}$

5. Which of the following statements describes a variable relationship that shows correlation but not causation?
   a. An elementary student's weight and the size of their vocabulary.
   b. The speed of a vehicle and the amount of time it takes to drive 60 miles with no traffic.
   c. The amount of time spent exercising and the number of calories burned.
   d. The dimensions of a rectangular garden and the amount of square footage available for planting.

6. Juan bought a pack of gum at the corner store. He gave 13 pieces of gum to his friend and has 11 pieces left. How many pieces of gum did the pack originally contain?
   a. 14 pieces of gum
   b. 18 pieces of gum
   c. 22 pieces of gum
   d. 24 pieces of gum

7. Which choice is NOT an example of the distributive property?
   a. $8(x - 5) = 8x - 40$
   b. $-2(y - 1) = -2y + 2$
   c. $3(4 + x) = 12 + 3x$
   d. $6(9 - y) = 54 + 9y$

8. Write a linear equation in standard form that represents the graph shown below.

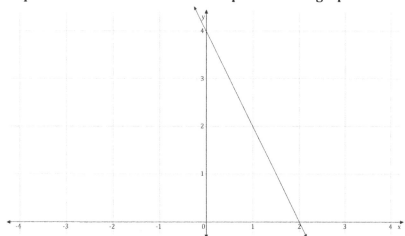

   a. $2x + 4 = y$
   b. $4x + y = 2$
   c. $x + 2y = 4$
   d. $2x + y = 4$

9. The square of 7 more than twice some number is 169. Which of the following could be the values of the number?
   a. −3 or 10
   b. −10 or 3
   c. −13 or 13
   d. −9 or 9

10. Solve the following quadratic using the complete the square method: $x^2 - 12x + 35 = 0$.

  a. $x = -2$ or $x = 6$
  b. $x = 2$ or $x = 6$
  c. $x = 5$ or $x = 7$
  d. $x = -7$ or $x = -5$

11. Poppy's Pizza charges $10 for each large pizza, plus a delivery fee of $3. The graph of this linear function passes through point $(3, 33)$ and has a slope of $m = 10$. Which graph shows the total cost for an order at Poppy's Pizza in relation to the number of pizzas ordered?

a.

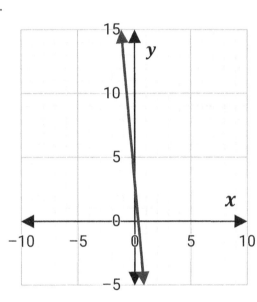

c.

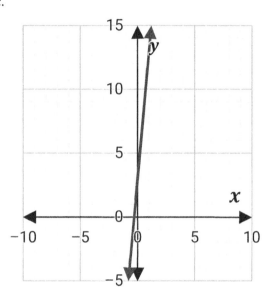

b.

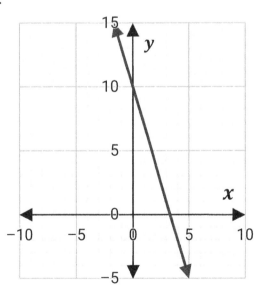

d.

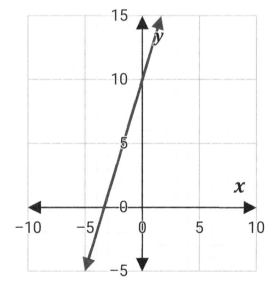

12. A rectangular pen has an area of $20x^2 + 55x$ square feet. The width of the pen is $5x$ feet. Find the length of the pen.
    a. $100x^3 + 275x^2$ feet
    b. $59x$ feet
    c. $4x + 11$ feet
    d. $4x^2 + 55$ feet

13. Simplify the expression: $\sqrt{50} + 5\sqrt{18}$.
    a. $15\sqrt{2}$
    b. $20\sqrt{2}$
    c. $5\sqrt{2}$
    d. $5\sqrt{68}$

14. Seven times a number x is less than 56. What values for the number make this true?
    a. $x < 8$
    b. $x \leq 8$
    c. $x \geq 8$
    d. $x > 8$

15. Graph the inequality: $y > -x + 1$.

a.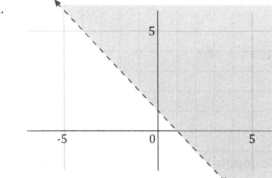

c.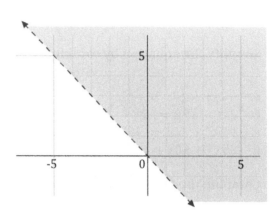

b.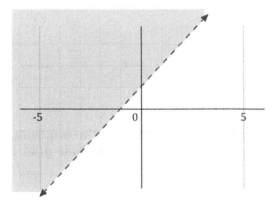

d.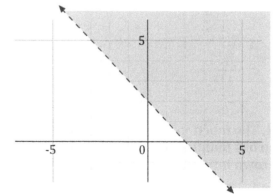

16. You have a square garden plot whose length and width are x meters. If you expand this into a rectangular plot by making it 6 meters longer and 4 meters wider, what will the area of the expanded plot be?

 a. $24x^2$ square meters
 b. $x^2 + 24x + 10$ square meters
 c. $x^2 + 10x + 24$ square meters
 d. $x^2 + 24$ square meters

17. Subtract $4xy - 8x^2 + 6$ from $2x^2 - 6xy + 4$.

 a. $2x^2 - 2x + 2$
 b. $10x^2 - 10xy - 2$
 c. $2x^2 - 12xy + 4$
 d. $10x^2 - 2 + 4$

18. Solve the following system of equations using the graphing method.
$$\begin{cases} 3x - 2y = -2 \\ x + 2y = 18 \end{cases}$$

 a. $(2, 8)$
 b. $(-2, -2)$
 c. $(-4, -5)$
 d. $(4, 7)$

19. Simplify the equation by combining like terms.
$$9x^2 - 6x - 11x - 16x^2 = -62$$

 a. $-7x^2 - 17x = -62$
 b. $7x^2 - 17x = 62$
 c. $-7x^2 - 10x = -62$
 d. $25x^2 + 10x = 62$

20. Combine the following algebraic fractions, and simplify the answer if possible: $\frac{10x+3}{2x+6} - \frac{4x-5}{2x+6}$.

 a. $\frac{-x+15}{2x+6}$
 b. $\frac{5x-1}{2x+6}$
 c. $\frac{3x+4}{x+3}$
 d. $\frac{x-1}{x+3}$

21. The function $f(x) = -8x^2 + 32x + 32$ models the height of a ball when thrown from the top of a building that is 32 feet high. What is the function of the height of the ball, $g(x)$, if it is thrown into the air from the ground floor?

 a. $g(x) = -8x^2 + 32x$
 b. $g(x) = -8x^2 + 32x + 15$
 c. $g(x) = -28x^2 + 52x + 15$
 d. $g(x) = -8x^2 + 32$

22. Evaluate the product: $(-10ab^5)\left(-\frac{1}{2}a^8b\right)$.
    a. $-5a^8b^5$
    b. $5a^8b^5$
    c. $-5a^9b^6$
    d. $5a^9b^6$

23. To better understand their shoppers and increase sales, Sasha's Boutique surveyed its customer base to find out the most common ages of the people who shop there. Many of their shoppers were one of two ages, which are represented by x in the equation $|x - 23| = 3$. Solve for x to determine the two ages.
    a. $x = 20, 23$
    b. $x = 20$
    c. $x = 20, 26$
    d. $x = 26$

24. Which choice shows an incorrect match between the inequality and the written description?
    a. $x > 3$: "x is greater than 3"
    b. $x < 3$: "x is less than 3"
    c. $x \geq 3$: "x is greater than or equal to 3"
    d. $x \leq 3$: "x is greater than or equal to 3"

25. Factor completely: $3a^2b^3 + 12ab^5$.
    a. $ab^3(3a + 12b^2)$
    b. $ab^3(3a + 12ab^2)$
    c. $3ab^3(a + 4b^2)$
    d. $3b^3(a^2 + 4ab^2)$

26. The height of the cone is represented by the variable h. If $h = \sqrt{12}$, what is the height of the cone in simplest form?
    a. $2\sqrt{3}$
    b. $3\sqrt{3}$
    c. $2\sqrt{5}$
    d. $12\sqrt{3}$

27. Simplify $-5a^2(a^3 - 2a^2 + 9a - 3)$.
    a. $-4a^6 - 7a^4 + 4a^2 - 8a^2$
    b. $-4a^5 - 7a^4 + 4a^3 - 8a^2$
    c. $-5a^6 + 10a^4 - 45a^2 + 15a^2$
    d. $-5a^5 + 10a^4 - 45a^3 + 15a^2$

28. Simplify the following expression:
$$(6b^2 + 4b + 2d) - (4b^2 + 18b + d)$$
    a. $10b^2 + 14b + 3d$
    b. $10b^2 + 14b + d$
    c. $2b^2 - 14b + d$
    d. $2b^2 - 14b + 3d$

29. A teacher gives a 100-point test to a large class. The student with the lowest score gets 30 points and the student with the highest score gets 80 points. The teacher considers two schemes for raising the scores by "curving" them. The first scheme raises the 30 to 40, the 80 to 100, and other scores to fit the line between these two extremes. The second scheme raises the 30 to 60, the 80 to 90, and other scores to fit the line between these extremes. Let x be a student's original score and y be the corresponding curved score. Write a system of equations that models these two schemes.

a. $\begin{cases} 6x - 5y = -20 \\ 3x - 5y = -210 \end{cases}$
b. $\begin{cases} 6x - 5y = -20 \\ 5x - 3y = -30 \end{cases}$
c. $\begin{cases} 5x - 6y = -90 \\ 3x - 5y = -210 \end{cases}$
d. $\begin{cases} 5x - 6y = -90 \\ 5x - 3y = -30 \end{cases}$

30. Solve the quadratic equation by using square roots: $(3x + 6)^2 - 144 = 0$.

a. $x = -12$ or $x = -12$
b. $x = -6$ or $x = 8$
c. $x = -6$ or $x = 2$
d. $x = -2$ or $x = 6$

31. The weight of an object on Earth varies directly as the same object's weight on the Moon. A 100-pound object on Earth weighs 16.5 pounds on the Moon. How much would an object weighing 20 pounds on Earth weigh on the Moon? Round your answer to the nearest tenth of a pound.

a. 13.9 pounds
b. 121.2 pounds
c. 3.3 pounds
d. 0.3 pounds

32. Factor $x^2 - 16x + 64$.

a. $(x - 16)(x + 4)$
b. $(x + 16)(x - 4)$
c. $(x - 8)^2$
d. $(x + 8)^2$

33. Write a function f in standard form, given that f is a quadratic function where $f(-4) = 0$, $f(2) = 0$, and $f(3) = -7$.

a. $f(x) = -x^2 - 2x + 8$
b. $f(x) = x^2 + 2x - 8$
c. $f(x) = \frac{1}{9}x^2 - 2x + 8$
d. $f(x) = \frac{1}{9}x^2 - \frac{2}{9}x - \frac{8}{9}$

34. Kristen is walking up a hill that is approximately 400 feet tall. She gains a constant elevation of 5 feet per second as she climbs the hill. Use a linear function to calculate the number of seconds it will take Kristen to reach the top of the hill.
    a. It will take Kristen 44 seconds to climb the hill.
    b. It will take Kristen 56 seconds to climb the hill.
    c. It will take Kristen 66 seconds to climb the hill.
    d. It will take Kristen 80 seconds to climb the hill.

35. Find the value of d in the equation $\frac{d}{3} = 14$.
    a. $d = 14$
    b. $d = 17$
    c. $d = 24$
    d. $d = 42$

36. Which graph correctly shows the solution set for $|x + 3| > 7$?

a.
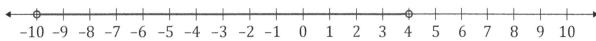

b.

c.
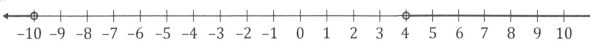

d.

37. Given the data below, which statement best describes the correlation coefficient (r), rounded to the nearest hundredth, of the number of police cars on a turnpike and the average speed of truck drivers?

Number of Patrol Cars	Average Speed (miles per hour)
3	63
4	62
5	59
6	52
7	55
8	51

   a. There is a strong negative linear association between the two variables.
   b. There is a moderate negative linear association between the two variables.
   c. There is a strong positive linear association between the two variables.
   d. There is no association between the two variables.

38. Solve the inequality: $-2x + 5 < 1$ or $x - 3 \geq 5$.
   a. $x < 2$ or $x \geq 8$
   b. $x > 2$
   c. $x \geq 8$
   d. $2 < x \leq 8$

39. The total revenue for selling a television can be represented by the expression $\frac{3n^2+5n+2}{2n-3}$. If the number of televisions sold can be expressed by $\frac{n^2+2n+1}{8n-12}$, which of the following expressions represents the unit price for a television set in terms of n?
   a. $\frac{n+1}{12n+8}$
   b. $\frac{12n+4}{n+1}$
   c. $\frac{12n+8}{2n+1}$
   d. $\frac{12n+8}{n+1}$

40. Solve the equation: $-2 + \sqrt{4x} = 6$.
   a. $x = 21$
   b. $x = 24$
   c. $x = 16$
   d. $x = 12$

41. Lamar and Ashley are playing a math game. One player needs to write a linear equation, and the other player needs to write an equation for a line that is perpendicular to the first one. Lamar starts by writing the equation $f(x) = -3x + 4$. If Ashley needs to write an equation for a line that is perpendicular to Lamar's, which equation could she write?
   a. $g(x) = -\frac{1}{3}x + 4$
   b. $g(x) = -4x + 3$
   c. $g(x) = -3x + 8$
   d. $g(x) = \frac{1}{3}x + 2$

42. The sum of one-half a number, t, and one-fourth of two more than the number is 20. What is the value of t?
   a. $t = 26$
   b. $t = 15$
   c. $t = \frac{26}{2}$
   d. $t = \frac{15}{2}$

43. Find the value of y in the equation $21 - 8y = 7y - 9$.
   a. $y = -30$
   b. $y = 30$
   c. $y = 2$
   d. $y = -2$

44. A fence is being built in your neighborhood park. One-third of the fence has been constructed. If the completed fence will be $24\sqrt{12}$ feet, how much of the fence has been built?

a. $9\sqrt{3}$ feet
b. $16\sqrt{3}$ feet
c. $16\sqrt{12}$ feet
d. $8\sqrt{4}$ feet

45. Determine whether the linear equations below are parallel. If so, what is their slope?

$$16x + 8y = 24$$
$$2x + y = 7$$

a. These lines are not parallel because their slopes are different.
b. These lines are not parallel because the slope of the line $2x + y = 7$ is undefined.
c. These lines are parallel. The slope of both lines is 2.
d. These lines are parallel. The slope of both lines is –2.

46. Use the distributive property to simplify the expression $3(b + 5) - 11$.

a. $3b - 6$
b. $3b + 4$
c. $15b - 11$
d. $15b + 4$

47. What is the graph of the function $y = -x^2 + 3x + 10$?

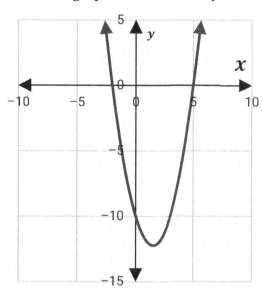

a.

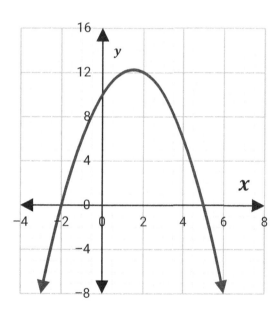

b.

c.

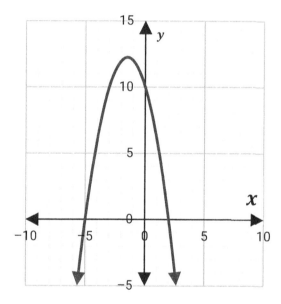

d.

48. A ball is thrown from 10 feet above the ground in the air with an initial velocity of 25 feet per second. The height, h, of the ball above the ground can be modeled by the quadratic equation $h = -16t^2 + 25t + 10$ where t is time, in seconds. To the nearest tenth of a second, how long will it take the ball to hit ground?
  a. 0.5 seconds
  b. 2.1 seconds
  c. 1.9 seconds
  d. 0.3 seconds

49. The volume of a rectangular prism with a square base can be expressed by the polynomial $3n^3 + 24n^2 + 48n$. What are the lengths of the sides of the square base and the height of the prism in terms of n?

   a. The lengths of the sides of the square base are $3n + 12$, and the height of prism is n.
   b. The lengths of the sides of the square base are $n^2 + 12$, and the height of prism is 3.
   c. The lengths of the sides of the square base are $n + 4$, and the height of prism is $3n$.
   d. The lengths of the sides of the square base are n, and the height of prism is $n + 4$.

50. Which of the following is an equation of a linear function in point-slope form that contains the points $(3, -6)$ and $(4, -3)$?

   a. $3x + 4y = 12$
   b. $y + 4 = 3(x + 3)$
   c. $y = 3x - 9$
   d. $y + 3 = 3(x - 4)$

51. Simplify the following algebraic expression. Leave your answer in factored form.
$$\frac{10x^2 - 5x}{15x}$$

   a. $\frac{2x^2 - 1}{3x}$
   b. $\frac{2x - 1}{3}$
   c. $\frac{1}{3}$
   d. $\frac{x}{3}$

52. The length of a rectangle is represented by the expression $\frac{10}{n}$ and the width by the expression $\frac{5}{4n}$. Which of the following expressions represents the perimeter of the rectangle?

   a. $\frac{15}{4n^3}$
   b. $\frac{15}{4n^2}$
   c. $\frac{90}{n}$
   d. $\frac{45}{2n}$

53. At a baseball game, Jake's concession stand sells hotdogs for $3 and sodas for $1.50. At the end of the night, Jake earned 300 dollars in sales. The equation $3x + 1.50y = 300$ can be used to determine the number of hotdogs and sodas sold in one night. In this equation, x represents the number of hotdogs sold, and y represents the number of sodas sold at the baseball game. Convert the equation from standard form to slope-intercept form.

   a. $y = -2x + 200$
   b. $y = -3x + 300$
   c. $y = 2x - 200$
   d. $y = 3x - 300$

54. Divide $2a^2 + 3a + 1$ by $a - 2$.
   a. $2a + 7 + \frac{15}{a-2}$
   b. $2a + 22$
   c. $2a + 7 - \frac{13}{a-2}$
   d. $2a - 6$

55. Mia opened a bank account in January and deposited $10. Each month, she plans to double the amount deposited in the previous month. Based on this information, how much money will Mia deposit in August?
   a. $1,380
   b. $1,280
   c. $138
   d. $128

56. What is the equation of a linear function in standard form that contains the points $(-1, 3)$ and $(4, -2)$?
   a. $x - y = 2$
   b. $x + y = -2$
   c. $-x + 2 = y$
   d. $x + y = 2$

57. You have just turned onto a highway. The table shows the mile marker M you can reach without speeding after t hours of driving at a constant rate under optimal traffic conditions. Write a linear inequality that represents the possible values of t and M under all traffic conditions.

t	1	2	3	5
M	73	133	193	313

   a. $60t + M \leq 133$
   b. $60t - M \leq -13$
   c. $60t + M \geq 133$
   d. $60t - M \geq -13$

58. Use the discriminant to determine how many times the graph of $y = x^2 + 2x - 8$ crosses the x-axis on the coordinate plane.
   a. The graph will touch the x-axis twice because the discriminant is positive.
   b. The graph will touch the x-axis once because the discriminant is 0.
   c. The graph does not touch the x-axis because the discriminant is negative.
   d. It cannot be determined how many times the graph crosses the x-axis using the discriminant.

59. State the range for the quadratic function $y = x^2 - 2x - 3$.
   a. $-\infty < y < \infty$
   b. $y \geq 1$
   c. $y \geq -4$
   d. $y \leq -4$

60. Factor $16s^2 + 121$.
   a. $(4s - 11)(4s + 11)$
   b. $(4s + 11)^2$
   c. $(4s - 11)^2$
   d. Cannot be factored

61. Which equation of a line contains point $(9, -1)$ and is perpendicular to the graph of the line $x + 3y = 6$?
   a. $y = 3x - 28$
   b. $y = 3x + 26$
   c. $y = -\frac{1}{3}x - 28$
   d. $y = -\frac{1}{3} + 26$

62. Factor $3x^2 + 11x + 10$.
   a. $(x - 1)(3x + 11)$
   b. $(x + 1)(3x + 11)$
   c. $(x + 5)(3x + 2)$
   d. $(x + 2)(3x + 5)$

63. Solve the following inequality: $\frac{x}{2} > 3$.
   a. $x < 6$
   b. $x > 6$
   c. $x \leq 6$
   d. $x \geq 6$

64. Which equation of a line is perpendicular to the x-axis, and what is its slope?
   a. $y = 8; m = 0$
   b. $y = 8; m = $ undefined
   c. $x = 8; m = 0$
   d. $x = 8; m = $ undefined

65. The length of a rectangle is $x + 4$, and the width is $x + 2$. If the area of the rectangle is 63 square feet, what is the value for x?
   a. $x = 4$
   b. $x = 7$
   c. $x = 5$
   d. $x = 2$

66. If $f(x) = x^2 - 3x$, then find the value of $\frac{f(8)}{f(2)}$.
   a. 20
   b. −20
   c. 4
   d. −4

67. Find the value of a in the equation $7a - 21 = 28$.
   a. $a = 49$
   b. $a = 42$
   c. $a = 1$
   d. $a = 7$

68. Aaron is looking for a new apartment for himself and his cat, Omar. He's narrowed his search down to two options. Apartment A costs $1,200 each month and has a one-time pet deposit of $300. Apartment B costs $1,050 each month and has a one-time pet deposit of $600. How many months will it take for the total costs of both apartments to be the same? Write and solve an equation using the variable m to represent months.
   a. 4 months
   b. 3 months
   c. 2 months
   d. 1 month

69. Sue is at the pet store looking for her dog's favorite brand of dog kibble. She has $80 to spend at the store, and she needs to buy 5 bags of food in order to stock up for a few months. Which of the following statements is true about the bags of food Sue can buy?
   a. Each bag of food needs to be less than or equal to $3.
   b. Each bag of food needs to be less than or equal to $3.20.
   c. Each bag of food needs to be less than or equal to $16.
   d. Each bag of food needs to be less than or equal to $5.

70. What is the value of y for the linear function $y = 5x + 3$ when the value of x is 8?
   a. 11
   b. 16
   c. 40
   d. 43

71. Convert $y = 3(x - 2)^2 + 1$ into standard form.
   a. $y = 3(x - 2)^2$
   b. $y = 3(x + 1)(x + 13)$
   c. $y = 3x^2 - 12x - 11$
   d. $y = 3x^2 - 12x + 13$

72. From the ground, a projectile is launched into the air. The projectile reaches its maximum height of 126 feet after 3 seconds. If the projectile follows a parabolic path, which of the following is the equation for the path of the projectile in terms of its height, h in feet, from the ground and time, t in seconds. State your answer in vertex form.

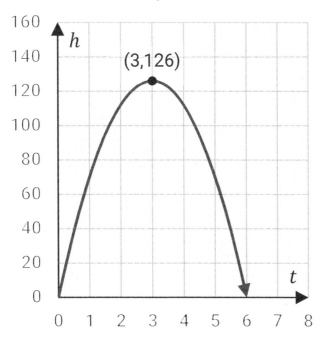

a. $h = -14(t-3)^2 + 126$
b. $h = 14(t-3)^2 + 126$
c. $h = -\frac{3}{126}(t-126)^2 + 3$
d. $h = -(t-126)^2 + 3$

73. Kayla goes for a walk every morning. The distance Kayla walks can be modeled by the equation $d = 3h$, where d is the distance walked in miles and h is the number of hours walked. Which graph best represents the relationship between x hours and y distance?

a.

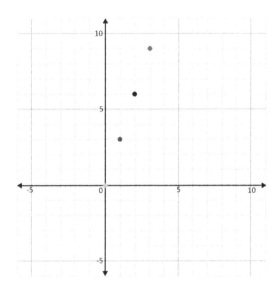

c.

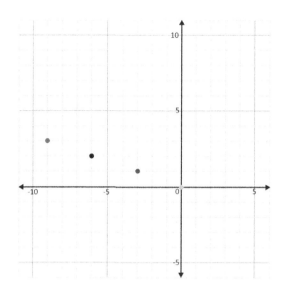

b.

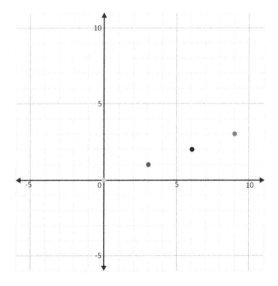

d.

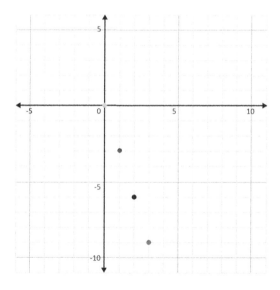

74. Solve the equation $-x + 8y = -14$ for y.
   a. $y = \frac{8}{-14-x}$
   b. $y = \frac{-14-x}{8}$
   c. $y = \frac{-14+x}{8}$
   d. $y = \frac{8}{-14+x}$

75. The product of three times the square of a number n and the sum of four times its cube and one. Which of the following represents the product in terms of n?
   a. $12n^6 + 43n^2$
   b. $7n^6 + 4n^2$
   c. $12n^5 + 3n^2$
   d. $7n^5 + 4n^3$

Answer Key and Explanations

1. B: To solve the inequality, we must isolate the variable on one side of the inequality. In this problem, we can start by applying the distributive property on the left side of the equation, multiplying the -2 by the $3x$ and then by the -5.

$$-6x + 10 \geq 58$$

The next step is to subtract 10 from both sides.

$$-6x \geq 48$$

Finally, we divide both sides by -6. Because we are dividing by a negative number, we need to "flip" the inequality.

$$x \leq -8$$

To graph $x \leq -8$, we start with a solid dot at -8 because our inequality is less than or equal to. The "or equal to" part tells us we need a solid dot. Then, because x is on the left, we follow the direction of the less than arrow by drawing our ray to the left. The arrow on the end of our ray indicates that all answers to the left on the number line satisfy the inequality.

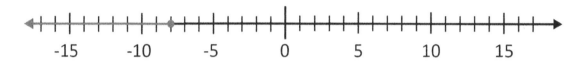

2. C: To find the value of x for the linear function when the value of y is 5, substitute 5 for y and evaluate the equation to find the value of x.

$$5 = 3x - 7$$

To solve for x, start by adding 7 to both sides.

$$12 = 3x$$

Then, divide both sides by 3.

$$x = 4$$

3. A: To rationalize the denominator, multiply by $\frac{\sqrt{42}}{\sqrt{42}}$.

$$\frac{30}{\sqrt{42}} = \frac{30}{\sqrt{42}} \cdot \frac{\sqrt{42}}{\sqrt{42}} = \frac{30\sqrt{42}}{42}$$

Then, simplify by dividing the common factor of 6 in the numerator and denominator

$$\frac{6 \cdot 5\sqrt{42}}{6 \cdot 7} = \frac{5\sqrt{42}}{7}$$

4. C: Use the rules for exponents to simplify.

$$\frac{9xy^5}{12x^4y^3} = \frac{9}{12} \cdot x^{1-4} \cdot y^{5-3}$$

Rewrite the expression using the rule $\frac{a^m}{a^n} = a^{m-n}$.

$$= \frac{3}{4} \cdot x^{-3} \cdot y^2$$

Reduce $\frac{9}{12}$ to $\frac{3}{4}$ and subtract the exponents for x and y.

$$= \frac{3y^2}{4x^3}$$

Rewrite the answer with no negative exponents using the rule $a^{-n} = \frac{1}{a^n}$.

5. A: Correlation refers to the strength of a linear relationship between two variables. Causation refers to instances in which one variable directly causes the outcome of another. In other words, correlation means that there is a relationship or connection between two variables, whereas causation means that one event explicitly causes another event to occur. Correlation and causation can exist at the same time, but correlation does not imply causation. In this question, we need to identify which variable relationship demonstrates correlation but not causation. There is no causation in the relationship between an elementary student's weight and the size of their vocabulary. While the two variables are correlated, one does not cause the other to happen. Therefore, the correct answer is A.

6. D: To start, analyze the problem and write an equation that represents the scenario. Since we don't know how many pieces of gum Juan started with, use a variable to represent this quantity. In this explanation, the variable g is used to represent the amount of gum Juan started with.

$$g$$

Since he gave 13 pieces of gum to his friend, subtract 13 from the original amount, g.

$$g - 13$$

Juan has 11 pieces of gum left, so set the expression equal to 11.

$$g - 13 = 11$$

To find the answer, solve for g in the equation. First, isolate the variable by doing inverse operations. Since the opposite of -13 is $+13$, add 13 to both sides of the equation.

$$g - 13 + 13 = 11 + 13$$

Combine like terms.

$$g = 24$$

To check your work, substitute the solution back into the original equation.

$$(24) - 13 = 11$$
$$11 = 11 ✓$$

7. D: When $6(9 - y)$ is multiplied using the distributive property, it becomes $54 - 6y$, not $54 + 9y$.

8. D: First, identify the y-intercept. The line intersects the y-axis at point $(0,4)$, so the y-intercept, b, of this linear equation is 4. From here, we need to identify the slope of the line, m. Use another point from the graph to find the slope. In this explanation, we will use point $(1,2)$ as well as the y-intercept, point $(0,4)$. To find the slope when given the coordinates of two points, find the change in y-values over the change in x-values. To do so, use the slope formula $m = \frac{y_2 - y_1}{x_2 - x_1}$. Substitute the x- and y-values from the two coordinates given.

$$m = \frac{4-2}{0-1} = \frac{2}{-1} = -2$$

The slope of this line is -2.

Since we now know the slope and the y-intercept, we can write an equation in slope–intercept form. The slope–intercept form of the equation is $y = mx + b$, where m is the slope, and b is the y-intercept. The slope-intercept equation of the line is $y = -2x + 4$.

Now that the equation is in slope–intercept form, change it into standard form, which is $Ax + By = C$. Move the x-term from the right side of the equation to the left by adding $2x$ to both sides of the equation. Therefore, the linear equation in two variables is $2x + y = 4$.

9. B: Let x represent the number. Then, the square of 7 more than twice x is represented by the expression $(7 + 2x)^2$. Since this expression equals to 169, we have the following quadratic equation.

$$(7 + 2x)^2 = 169$$

Take the square root of both sides to solve for x.

$$(7 + 2x)^2 = 169$$
$$\sqrt{(7 + 2x)^2} = \sqrt{169}$$
$$7 + 2x = \pm 13$$
$$2x = -7 \pm 13$$
$$\frac{2x}{2} = \frac{-7 \pm 13}{2}$$
$$x = \frac{-7 - 13}{2} \text{ or } x = \frac{-7 + 13}{2}$$
$$x = \frac{-20}{2} \text{ or } x = \frac{6}{2}$$
$$x = -10 \text{ or } x = 3$$

So, x can be either -10 or 3.

10. C: First, complete the square to put the equation into the form $(ax + b)^2 = c$.

$x^2 - 12x = -35$	Move the constant term to the right-hand side of the equation.
$x^2 - 12x = -35$	Locate the coefficient to the linear term.
$\left(\frac{-12}{2}\right)^2 = (-6)^2 = 36$	Take one-half of this coefficient and square it.

$$x^2 - 12x + 36 = -35 + 36 \qquad \text{Add the squared value to both sides of the equation.}$$

$$(x - 6)^2 = 1 \qquad \text{Factor the perfect square trinomial } x^2 - 12x + 36.$$

Since the converted equation is in the form $(ax + b)^2 = c$, we can take the square root of both sides to solve for x.

$$(x - 6)^2 = 1$$
$$\sqrt{(x - 6)^2} = \sqrt{1}$$
$$x - 6 = \pm 1$$
$$x - 6 + 6 = 6 \pm 1$$
$$x = 6 \pm 1$$
$$x = 6 - 1 \text{ or } x = 6 + 1$$
$$x = 5 \text{ or } x = 7$$

11. C: To graph this linear function given the slope and a point the line passes through, determine the slope-intercept equation of the line. The slope-intercept equation of a line is $y = mx + b$, where m stands for the slope, and b stands for the y-intercept. Since we know the slope, m, is 10, start by replacing m with 10 in the equation.

$$y = 10x + b$$

Next, substitute the x- and y-coordinates into the equation. Since the y-coordinate is 33, replace y with 33 in the equation. Since the x-coordinate is 3, replace x with 3 in the equation.

$$(33) = 10(3) + b$$

From here, solve the equation for b to identify the y-intercept. Simplify the equation by multiplying 10 and 3.

$$33 = 30 + b$$

Next, solve for b by subtracting 30 from both sides.

$$b = 3$$

Finally, substitute 3 into the equation for b. The slope-intercept equation of a line that passes through point (3,33) and has a slope of $m = 10$ is $y = 10x + 3$.

Now that the linear equation is in slope-intercept form, graph the function. Since the y-intercept is 3, the line will intersect the y-axis at 3, or (0,3). Plot a point at this location. From here, use the slope to move to the next point. Since the slope is 10, move 10 units up and 1 unit right to find the next point on the graph. The coordinate pair for this point is (1,13). Repeat this process to continue plotting points on the line of this linear function. When the value of x is 3, the value of y will be 33. The graph in choice C shows all these points on the line.

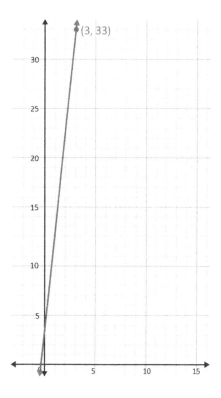

12. C: The area of a rectangle is the product of its length and width, $A = lw$. We can rearrange this to calculate the length as the area divided by the width, $l = \frac{A}{w}$. The area is $20x^2 + 55x$ square feet and the width is $5x$ feet. Substitute the expressions into the formula.

$$l = \frac{A}{w} = \frac{20x^2 + 55x}{5x}$$

Next, we separate the fraction into two fractions, putting each term of the numerator over the denominator. Then we divide coefficients by coefficients and variables by variables, recalling that $\frac{x^m}{x^n} = x^{m-n}$ and that a fraction with the same numerator and denominator equals 1.

$$\frac{20x^2 + 55x}{5x} = \frac{20x^2}{5x} + \frac{55x}{5x} = \frac{20}{5} \cdot \frac{x^2}{x} + \frac{55}{5} \cdot \frac{x}{x} = 4x^{2-1} + 11 \cdot 1 = 4x + 11$$

Therefore, the length of the pen is $4x + 11$ feet.

13. B: Simplify the expression so that the radicals are the same.

$\sqrt{50}$ can be simplified to $\sqrt{25}\sqrt{2}$, or $5\sqrt{2}$.

$5\sqrt{18}$ can be simplified to $5\sqrt{9}\sqrt{2}$, or $15\sqrt{2}$.

Rewrite $\sqrt{50} + 5\sqrt{18}$ as $5\sqrt{2} + 15\sqrt{2}$. Now that the radicals are the same, the coefficients can be added.

$$5\sqrt{2} + 15\sqrt{2} = 20\sqrt{2}$$

14. A: To solve this problem, we must first translate the written problem into a math inequality. "Seven times a number x" translates to $7x$. "Is less than 56" translates to the less than symbol ($<$) and the constant 56.

$$7x < 56$$

To solve this, divide both sides of the inequality by 7.

$$x < 8$$

15. A: The inequality $y > -x + 1$ is in slope-intercept form. To graph the inequality, treat it as though it is an equation and identify the y-intercept and the slope. If we treat $y > -x + 1$ as though it is $y = -x + 1$, we know that the y-intercept is 1, and the slope is -1. Now the line can be graphed. The line will be dashed because the inequality does not include "equal to." The shaded region will be above the line because the inequality reads, "y is greater than," so shade the y-values greater than the line.

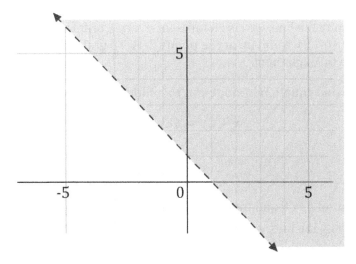

The shaded region above the dashed line means that any value in the red area is considered a solution for the inequality. The dashed line means that values on the line are not solutions.

16. C: Increasing the original length of x by 6 produces a new length of $(x + 6)$ meters. Similarly, the new width will be $(x + 4)$ meters. Since the area of a rectangle is the product of its length and width, The area of the new rectangle will be $(x + 6)(x + 4)$ square meters. Calculate this product using the FOIL method and combine like terms.

$$\begin{aligned}(x+6)(x+4) &= \overbrace{x \cdot x}^{F} + \overbrace{x \cdot 4}^{O} + \overbrace{6 \cdot x}^{I} + \overbrace{6 \cdot 4}^{L} \\ &= x^2 + 4x + 6x + 24 \\ &= x^2 + 10x + 24\end{aligned}$$

The area of the expanded plot will be $x^2 + 10x + 24$ square meters.

17. B: Polynomials can be subtracted vertically. Line up the like terms in the first polynomial under the second polynomial.

$$\begin{array}{rrrr} & 2x^2 & -6xy & +4 \\ - & (-8x^2 & +4xy & +6) \end{array}$$

Take the additive inverse of the bottom polynomial by distributing the negative sign to the bottom polynomial. Then remove the parentheses from both polynomials before combining them.

$$\begin{array}{rrrr} & 2x^2 & -6xy & +4 \\ + & 8x^2 & -4xy & -6 \\ \hline & 10x^2 & -10xy & -2 \end{array}$$

18. D: To solve this system by the graphing method, we graph both lines by any convenient method. For instance, we can solve the first equation for y, putting it in slope-intercept form, $y = mx + b$.

$$3x - 2y = -2$$
$$-2y = -3x - 2$$
$$y = \frac{3}{2}x + 1$$

The line has a y-intercept of $b = 1$ and therefore passes through the point $(0,1)$. The slope $m = \frac{3}{2}$ indicates that moving along the line, we rise 3 units for every 2 units we move to the right. If we do this starting at the point $(0,1)$, we end up at the point $(2,4)$. To graph the line, we plot these two points and draw the line through them.

Similarly, we put the second equation in slope-intercept form.

$$x + 2y = 18$$
$$2y = -x + 18$$
$$y = -\frac{1}{2}x + 9$$

The line has a y-intercept of $b = 9$ and therefore passes through the point $(0,9)$. The slope $m = -\frac{1}{2}$ indicates that moving along the line, we rise –1 unit (that is, we drop 1 unit) for every 2 units we move to the right. If we do this starting at the point (0,9), we end up at the point (2,8). Again, we plot the points and draw the line through them. Here is the graph of both lines.

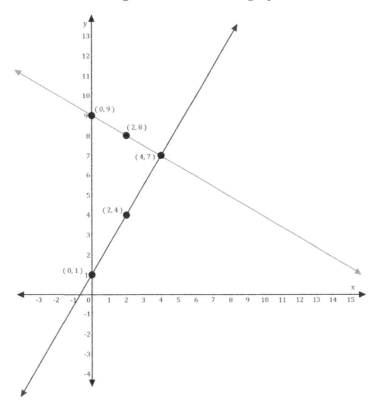

Since each line represents all the solutions to one equation, their intersection point—the point they have in common—solves both equations. That is, it solves the system. These lines intersect at the point (4,7).

19. A: To combine like terms, look for terms that have the same variable part. Once identified, combine them into a single term by adding or subtracting their coefficients.

$$9x^2 - 6x - 11x - 16x^2 = -62$$

Start by combining the terms $9x^2$ and $-16x^2$ on the left side of the equation.

$$-7x^2 - 6x - 11x = -62$$

Next, combine the terms $-6x$ and $-11x$ on the left side of the equation.

$$-7x^2 - 17x = -62$$

20. C: Since the denominators are the same, combine the numerators to write the expression as one fraction.

$$\frac{10x+3}{2x+6} - \frac{4x-5}{2x+6} = \frac{10x+3-(4x-5)}{2x+6} = \frac{6x+8}{2x+6}$$

Simplifying the answer by factoring out a 2 in both the numerator and denominator.

$$\frac{6x+8}{2x+6} = \frac{2(3x+4)}{2(x+3)} = \frac{3x+4}{x+3}$$

21. A: Since the ball will now be tossed from 32 feet below the original throw, the entire graph will move down by 32 units, which is the same as subtracting 32 from the original function.

$$g(x) = -8x^2 + 32x + 32 - 32$$
$$g(x) = -8x^2 + 32x$$

22. D: Using commutativity and associativity of multiplication and the law of exponents that states $x^m \cdot x^n = x^{m+n}$, we get $(-10ab^5)\left(-\frac{1}{2}a^8 b\right) = (-10)\left(-\frac{1}{2}\right) \cdot a \cdot a^8 \cdot b^5 \cdot b = 5a^{1+8}b^{5+1} = 5a^9 b^6$.

23. C: For this problem, we are given the equation $|x - 23| = 3$. Since the expression in absolute value bars containing x is already by itself on the left side, the first step is dropping the absolute value bars by writing $\pm$ in front of the 3.

$$x - 23 = \pm 3$$

Now, add 23 to both sides to get x by itself.

$$x = \pm 3 + 23$$

If 3 is positive, then $x = 3 + 23 = 26$, and if 3 is negative, then $x = -3 + 23 = 20$. Most of the shoppers at Sasha's Boutique are either 20 or 26 years old.

24. D: Choice D gives the inequality, $x \leq 3$. By definition, this inequality is read as, "x is less than or equal to 3." This contradicts the statement in the answer choice.

25. C: First, determine the greatest common factor (GCF) for the terms by decomposing each of them into their prime factors.

$$3a^2 b^3 = 3 \cdot a \cdot a \cdot b \cdot b \cdot b \text{ and } 12ab^5 = 2 \cdot 2 \cdot 3 \cdot a \cdot b \cdot b \cdot b \cdot b \cdot b$$

To find the GCF, compare common prime factors for each term (bolded below).

$$\mathbf{3} \cdot a \cdot \mathbf{a} \cdot \mathbf{b} \cdot \mathbf{b} \cdot \mathbf{b} \text{ and } 2 \cdot 2 \cdot \mathbf{3} \cdot \mathbf{a} \cdot \mathbf{b} \cdot \mathbf{b} \cdot \mathbf{b} \cdot b \cdot b$$

The GCF is the product of the common prime factors.

$$3 \cdot a \cdot a \cdot b \cdot b \cdot b = 3ab^3$$

Factoring the GCF from each term we have:

$$3a^2 b^3 + 12ab^5 = 3ab^3(a + 4b^2)$$

We can always check our answer by taking the product of the GCF and the expression inside the parentheses by the distribution property of multiplication.

$$3ab^3(a + 4b^2) = 3ab^3 \cdot a + 3ab^3 \cdot 4b^2 = 3a^2 b^3 + 12ab^5$$

26. A: $\sqrt{12}$ can be expressed as the product of two radicals: $\sqrt{4}\sqrt{3}$. The square root of 4 is 2, so $\sqrt{4}\sqrt{3}$ becomes $2\sqrt{3}$. The height of the cone in simplest form is $2\sqrt{3}$.

27. D: Use the distributive property for multiplication to simplify.

$$-5a^2(a^3 - 2a^2 + 9a - 3) = (-5a^2) \cdot a^3 - (-5a^2) \cdot 2a^2 + (-5a^2) \cdot 9a - (-5a^2) \cdot 3$$
$$= -5a^5 + 10a^4 - 45a^3 + 15a^2$$

28. C: Start by removing the parentheses from the first portion of the expression.

$$6b^2 + 4b + 2d - (4b^2 + 18b + d)$$

Next, use the distributive property to distribute the negative sign to each term in the second set of parentheses. To do so, multiply each term in parentheses by −1.

$$6b^2 + 4b + 2d - 4b^2 - 18b - d$$

From here, combine like terms. To do so, look for terms that have the same variable part. Once identified, combine them into a single term by adding or subtracting their coefficients. Start with $6b^2$ and $-4b^2$.

$$2b^2 + 4b + 2d - 18b - d$$

Next, combine the terms $4b$ and $-18b$.

$$2b^2 - 14b + 2d - d$$

Finally, combine the terms $2d$ and $-d$.

$$2b^2 - 14b + d$$

29. A: Under the first scheme, a score of 30 becomes 40 and a score of 80 becomes 100, so the points (30,40) and (80,100) define the line for the first scheme. Substituting these two points into the slope formula, we find the slope of this line:

$$m = \frac{y_2 - y_1}{x_2 - x_1} = \frac{100 - 40}{80 - 30} = \frac{60}{50} = \frac{6}{5}$$

Now we substitute this slope and one of the points, say (30,40), into the point-slope form of a linear equation, $y - y_1 = m(x - x_1)$, and simplify, getting an equation for the first scheme.

$$y - 40 = \frac{6}{5}(x - 30)$$
$$y - 40 = \frac{6}{5}x - 36$$
$$-\frac{6}{5}x + y = 4$$
$$6x - 5y = -20$$

Next, we repeat these steps for the second scheme. The points (30,60) and (80,90) define the line for this scheme. Substituting these two points into the slope formula, we find the slope of this line is:

$$m = \frac{90 - 60}{80 - 30} = \frac{30}{50} = \frac{3}{5}$$

Again, we substitute this slope and one of the points, say (30,60), into the point-slope form of a linear equation and simplify, getting an equation for the second scheme.

$$y - 60 = \frac{3}{5}(x - 30)$$
$$y - 60 = \frac{3}{5}x - 18$$
$$-\frac{3}{5}x + y = 42$$
$$3x - 5y = -210$$

The equations for the first and second schemes form the desired system of equations.

30. C: First, put the quadratic equation in the form $(ax + b)^2 = c$ by adding 144 to both sides.

$$(3x + 6)^2 - 144 = 0$$
$$(3x + 6)^2 = 144$$

Now, take the square root of both sides to solve for x.

$$3x + 6 = \pm 12$$
$$3x = -6 \pm 12$$
$$x = \frac{-6 - 12}{3} \text{ or } x = \frac{-6 + 12}{3}$$
$$x = \frac{-18}{3} \text{ or } x = \frac{6}{3}$$
$$x = -6 \text{ or } x = 2$$

31. C: Since the weight varies directly, set up the following proportion and solve for the weight of the object on the Moon.

$$\frac{100}{16.5} = \frac{20}{M}$$
$$330 = 100M$$
$$3.3 = M$$

Additionally, we can use the equation $y = kx$ to solve direct variation problems. This problem refers to the relationship between the weight of an object on Earth and the Moon. Instead of using the variables x and y, we'll use the variable E to represent the weight of the object on Earth, and the variable M to represent the weight of the object on the Moon.

$$y = kx \rightarrow E = kM$$

First, rewrite the equation by substituting 100 for E and 16.5 for M. Then, solve for k to find the constant variation.

$$100 = k(16.5)$$

$$\frac{100}{16.5} = \frac{16.5k}{16.5}$$

$$6.06 \approx k$$

Next, substitute 6.06 in the equation $E = kM$ for k.

$$E = 6.06\,M$$

To find the value of M when $E = 20$, substitute 20 into the equation for E. Then, solve for M.

$$20 = 6.06\,M$$
$$\frac{20}{6.06} = \frac{6.06M}{6.06}$$
$$M \approx 3.3$$

Therefore, an object weighing 20 pounds on Earth would weigh about 3.3 pounds on the moon.

32. C: A factorable quadratic trinomial in the form $a^2 - 2ab + b^2$ is called a perfect square trinomial, where a^2 and b^2 are perfect squares, and factor to $(a - b)(a - b) = (a - b)^2$.

For our trinomial, x^2 and 64 are perfect squares so, $a = x$ and $b = 8$. Since $2ab = 2x \cdot 8 = 16x$, we have the following factorization.

$$x^2 - 16x + 64 = (x - 8)^2$$

33. A: In factored form, a quadratic function having two real zeros can be written as $f(x) = a(x - m)(x - n)$, where m and n are the zeros of the function, and a is a real number that does not equal 0.

Since $f(-4) = 0$ and $f(2) = 0$, the values −4 and 2 are the zeros for the function. Substitute the zeros into the factored form equation.

$$f(x) = a\big(x - (-4)\big)(x - 2)$$
$$f(x) = a(x + 4)(x - 2)$$

Then, substitute $f(3) = -7$ and solve for a.

$$-7 = a(3 + 4)(3 - 2)$$
$$-7 = a(7)(1)$$
$$-7 = 7a$$
$$a = -1$$

Finally, substitute the value for a and expand the factored form equation to write the quadratic function in standard form.

$$f(x) = -1(x+4)(x-2)$$
$$= -1(x^2 - 2x + 4x - 8)$$
$$= -1(x^2 + 2x - 8)$$
$$= -x^2 - 2x + 8$$

So, the quadratic function is $f(x) = -x^2 - 2x + 8$.

34. D: A linear function can be written to describe Kristen's walk up the hill. Her elevation depends on how many seconds she has been walking, so the variable, x, will represent time in seconds, and the variable, y, will represent the elevation in feet, expressed as $y = 5x$. The hill is 400 feet, so $400 = 5x$ becomes $x = 80$. It will take Kristen 80 seconds to climb the hill.

35. D: Solve for d in the equation.

$$\frac{d}{3} = 14$$

Isolate the variable by doing inverse operations. Multiply both sides of the equation by 3.

$$d = 42$$

36. C: To solve, isolate the variable x in the inequality. Since this inequality involves absolute value, rewrite this inequality as two separate inequalities without absolute value bars. $|x + 3| > 7$ becomes $x + 3 > 7$ and $x + 3 < -7$. Notice that the sign is flipped in $x + 3 < -7$. From here, isolate x in the first inequality by doing inverse operations. Subtract 3 from both sides of the inequality.

$$x + 3 - 3 > 7 - 3$$
$$x > 4$$

Next, isolate x in the second inequality by doing inverse operations. Subtract 3 from both sides of the inequality.

$$x + 3 - 3 < -7 - 3$$
$$x < -10$$

From here, graph $x > 4$ or $x < -10$ on the same number line. The first inequality states that x is greater than 4, so the number line must show an open circle at 4 with a line extending to the right. The second inequality states that x is less than –10, so the number line must also show an open circle at –10 with a line extending to the left.

37. A: The correlation coefficient, represented by the variable r, is a measurement of the strength of the linear relationship between two quantitative variables or sets of data. It is a number ranging from –1.0 to 1.0, with 1.0 indicating a perfect positive linear association, 0 indicating no association, and –1.0 indicating a perfect negative linear association. In this scenario, the correlation coefficient represents the relationship between the number of patrol cars on the turnpike and the average speed of truck drivers. The correlation coefficient (r) is found by dividing the covariance by the product of the standard deviations of each variable. The correlation coefficient of this set of data can be calculated using a calculator and is approximately –0.92. Since –0.92 is a negative number, it indicates a negative association. Since –0.92 is very close to –1.0, a correlation coefficient of –0.92

indicates a strong negative linear association. When this data is plotted on the coordinate plane, it shows a nearly perfect straight line, pointing in a negative direction.

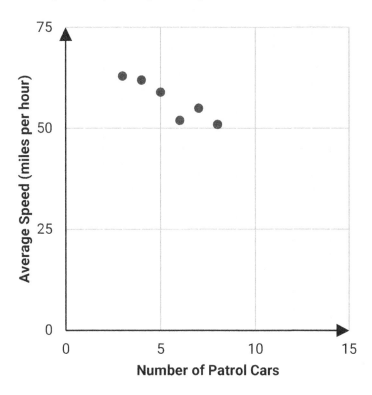

38. B: To solve an "or" inequality, we must solve each part of the inequality separately. We can solve the first part by subtracting 5 from both sides and then dividing both sides by –2.

$$-2x + 5 < 1$$
$$-2x < -4$$
$$x > 2$$

We can solve the second part by adding 3 to both sides.

$$x - 3 \geq 5$$
$$x \geq 8$$

Now we check to see if the two parts overlap. The answer can be expressed as $x > 2$, since any value that satisfies at least one of the two parts counts as a solution. We can see this illustrated on the number line below.

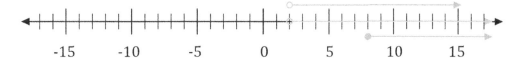

39. D: The unit price for the televisions can be found using the equation unit price = total revenue ÷ number of units sold. Substituting the given values, we have the following equation.

$$\text{unit price} = \frac{3n^2 + 5n + 2}{2n - 3} \div \frac{n^2 + 2n + 1}{8n - 12}$$

To divide, multiply by the reciprocal of the second fraction.

$$\text{unit price} = \frac{3n^2 + 5n + 2}{2n - 3} \cdot \frac{8n - 12}{n^2 + 2n + 1}$$

To simplify, first completely factor the numerator and denominator of each fraction.

$$\frac{(3n + 2)(n + 1)}{2n - 3} \cdot \frac{4(2n - 3)}{(n + 1)(n + 1)}$$

Next, cancel out common factors.

$$\frac{(3n + 2)\cancel{(n + 1)}}{\cancel{2n - 3}} \cdot \frac{4\cancel{(2n - 3)}}{\cancel{(n + 1)}(n + 1)}$$

Finally, multiply the remaining factors.

$$\frac{3n + 2}{1} \cdot \frac{4}{n + 1} = \frac{12n + 8}{n + 1}$$

In terms of n, the unit price of a television is $\frac{12n+8}{n+1}$.

40. C: Isolate the radical by adding 2 to each side.

$$-2 + 2 + \sqrt{4x} = 6 + 2$$
$$\sqrt{4x} = 8$$

Get rid of the radical by squaring both sides of the equation.

$$\left(\sqrt{4x}\right)^2 = (8)^2$$
$$4x = 64$$

Divide both sides by 4.

$$\frac{4x}{4} = \frac{64}{4}$$
$$x = 16$$

Since both sides of the equation were squared when solving, we need to check for extraneous solutions. However, since no solution is not an answer choice, you can assume $x = 16$ is not an extraneous solution.

41. D: The given function and answer choices are in the form $f(x) = mx + b$, where m is the slope and b is the y-intercept. Perpendicular lines have slopes that are the opposite reciprocal of each other. Start by identifying the slope, m, in the original equation, $f(x) = -3x + 4$. From looking at the equation, we see that the slope is –3. Next, identify the opposite reciprocal of –3. The opposite reciprocal of –3 is $\frac{1}{3}$. An equation for a line that is perpendicular to Lamar's line must have a slope of $\frac{1}{3}$. The equation $g(x) = \frac{1}{3}x + 2$ is the only answer option given with a slope of $\frac{1}{3}$.

42. A: One-half of t can be represented by the expression $\frac{t}{2}$, and one-fourth of two more than the number can be represented by $\frac{t+2}{4}$. Since the sum of the two expressions is 20, we have the following equation.

$$\frac{t}{2} + \frac{t+2}{4} = 20$$

To solve a rational equation, multiply both sides of the equation by the lowest common denominator (LCD), to clear out any denominators. The LCD for the denominators 2 and 4 is 4, so multiply both sides of the equation by 4.

$$4\left(\frac{t}{2} + \frac{t+2}{4}\right) = 4 \cdot 20$$
$$2t + t + 2 = 80$$

Now that the denominators are cleared out, solve the equation to find the value for t.

$$2t + t + 2 = 80$$
$$3t + 2 = 80$$
$$3t = 78$$
$$t = 26$$

43. C: Solve for y in the equation.

$$21 - 8y = 7y - 9$$

Since there are variables on both sides of the equal sign, move all variables to one side and combine like terms. Add $8y$ to both sides of the equation.

$$21 - 8y + 8y = 7y + 8y - 9$$
$$21 = 15y - 9$$

Next, add 9 to both sides of the equation.

$$21 + 9 = 15y - 9 + 9$$
$$30 = 15y$$

Finally, divide both sides of the equation by 15.

$$\frac{30}{15} = \frac{15y}{15}$$
$$2 = y$$

44. B: The total length of the fence will be $24\sqrt{12}$ feet. One-third of the fence has been completed, so divide $24\sqrt{12}$ by 3. $\frac{24\sqrt{12}}{3}$ can be simplified to $8\sqrt{12}$. The square root of 12 can be split into the factors 4 and 3, giving the expression $8\sqrt{4}\sqrt{3}$, and then $16\sqrt{3}$. Thus, $16\sqrt{3}$ feet of fence has been built.

45. D: Two lines are parallel if they have the same slope and different y-intercepts. To determine whether the linear equations are parallel, start by identifying the slope, m, of each line. To find the slope, write each equation in slope-intercept form, $y = mx + b$.

$$16x + 8y = 24$$

First, move the x-term to the right side of the equation by subtracting $16x$ from both sides of the equation.

$$16x - 16x + 8y = 24 - 16x$$
$$8y = 24 - 16x$$

Next, isolate the variable y by dividing both sides of the equation by 8.

$$\frac{8y}{8} = \frac{24 - 16x}{8}$$
$$y = 3 - 2x$$
$$y = -2x + 3$$

Since the value of m in the equation is –2, the slope of this line is –2.

Now, look at the second equation. Write this equation in slope-intercept form, $y = mx + b$.

$$2x + y = 7$$

First, move the x-term to the right side of the equation by subtracting $2x$ from both sides of the equation.

$$2x - 2x + y = 7 - 2x$$
$$y = 7 - 2x$$
$$y = -2x + 7$$

Since the value of m in the equation is –2, the slope of the line is –2.

Written in slope-intercept form, both equations have the same slope, −2, but different y-intercepts. Therefore, these two linear equations are parallel.

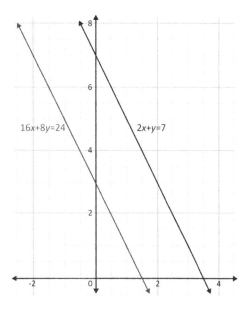

46. B: Apply the distributive property to the expression by multiplying each term inside of the parentheses by 3.

$$3(b + 5) - 11$$

$$3b + 15 - 11$$

Combine like terms to simplify the expression.

$$3b + 4$$

47. B: A quadratic function in standard form, $y = ax^2 + bx + c$, where a, b, and c are real numbers and $a \neq 0$, opens up when $a > 0$ and opens down when $a < 0$. Since $a = -1$ for our quadratic function, its graph opens down.

The x-coordinate for the vertex for a quadratic function can be found using the formula $x = -\frac{b}{2a}$. The y-coordinate for the vertex can be found by substituting $x = -\frac{b}{2a}$ into the function.

$$x = -\frac{3}{2(-1)} = -\frac{3}{-2} = \frac{3}{2} = 1.5$$

Therefore, the x-coordinate of the vertex is 1.5. Next, substitute $x = \frac{3}{2}$ into the function.

$$y = -\left(\frac{3}{2}\right)^2 + 3\left(\frac{3}{2}\right) + 10$$
$$y = -\frac{9}{4} + \frac{9}{2} + 10$$
$$y = \frac{49}{4}$$
$$y = 12.25$$

So, the vertex is $\left(\frac{3}{2}, \frac{49}{4}\right)$, or $(1.5, 12.25)$.

Then, locate additional points on the graph by identifying the x-intercepts of the graph. Substitute $y = 0$ into the equation to find the x-intercepts.

$$0 = -x^2 + 3x + 10$$
$$-1 \cdot 0 = -1(-x^2 + 3x + 10)$$
$$0 = x^2 - 3x - 10$$
$$0 = (x+2)(x-5)$$

$x + 2 = 0$ $x - 5 = 0$

$x = -2$ $x = 5$

So, the x-intercepts are $(-2, 0)$ and $(5, 0)$. Use all these points to find the graph of the function in choice B.

48. C: A quadratic equation can be solved for x when it is written in standard form by using the quadratic formula. The standard form equation of a quadratic is $ax^2 + bx + c = 0$, where a, b, and c are constants and $a \neq 0$.

The quadratic formula is $x = \frac{-b \pm \sqrt{b^2 - 4ac}}{2a}$.

To find the time when the ball hits the ground, set $h = 0$.

$$0 = -16t^2 + 25t + 10$$

Now the quadratic equation is in standard form where $a = -16$, $b = 25$, and $c = 10$. Substituting our values into the quadratic formula and solving for t, we get the following equation.

$$t = \frac{-(25) \pm \sqrt{(25)^2 - 4(-16)(10)}}{2(-16)}$$
$$= \frac{-25 \pm \sqrt{625 + 640}}{-32}$$
$$= \frac{-25 \pm \sqrt{1{,}265}}{-32}$$
$$= \frac{25 \pm \sqrt{1{,}265}}{32}$$

We can separate our two solutions as follows.

$$t = \frac{25\sqrt{1{,}265}}{32} = -0.3 \text{ or } t = \frac{25 + \sqrt{1{,}265}}{32} = 1.9$$

While both solutions satisfy the quadratic equation, time cannot have a negative value so we can disregard our answer of −0.3. Thus, to the nearest tenth of a second, it takes 1.9 seconds for the ball to hit the ground.

49. C: The volume, V, of a rectangular prism with a square base is $V = s^2 h$, where s is the length of a side of the square and h is the height of the prism.

Completely factor the polynomial.

$3n^3 + 24n^2 + 48n = 3n(n^2 + 8n + 16)$	Factor out the greatest common factor (GCF) of $3n$ from each term in the polynomial.
$\begin{aligned}3n(n^2 + 8n + 16) &= 3n(n^2 + 2 \cdot n \cdot 4 + 4^2) \\ &= 3n(n+4)^2\end{aligned}$	The quadratic trinomial is a perfect square trinomial that can be factored as $a^2 + 2ab + b^2 = (a+b)^2$.

So, the factored form for the volume of the prism can be modeled by $V = 3n(n+4)^2$. Thus, $h = 3n$, and $s = n + 4$.

50. D: The point-slope equation of a line is $y - y_1 = m(x - x_1)$, where x_1 and y_1 are the coordinates of a point on the line (x_1, y_1), and m is the slope. Start by identifying the slope using the slope formula $m = \frac{y_2 - y_1}{x_2 - x_1}$. Substitute the x- and y-values from the given coordinates.

$$m = \frac{y_2 - y_1}{x_2 - x_1}$$
$$m = \frac{-3 - (-6)}{4 - 3}$$
$$m = \frac{3}{1} = 3$$

Since we know the slope is 3, replace m with 3 in the equation.

$$y - y_1 = m(x - x_1)$$
$$y - y_1 = 3(x - x_1)$$

Next, substitute the x- and y-coordinates from one of the points into the equation. In this explanation, we will use point $(4, -3)$.

$$y - (-3) = 3(x - 4)$$

51. B: Factor the numerator and denominator completely.

$$\frac{10x^2 - 5x}{15x} = \frac{5x(2x - 1)}{5(3x)}$$

Divide common factors from the numerator and denominator.

$$\frac{\cancel{5}\cancel{x}(2x-1)}{\cancel{5}(3\cancel{x})} = \frac{2x-1}{3}$$

52. D: The perimeter of a rectangle can be found by taking twice the sum of the length and width.

$$P = 2(l+w)$$
$$P = 2\left(\frac{10}{n} + \frac{5}{4n}\right)$$

To combine two rational expressions with unlike denominators, we must first find the least common denominator (LCD). The LCD is the lowest common multiple between the two denominators, which in this case is $4n$. Multiply the first expression in the parentheses by some form of 1 that will give it the LCD.

$$2\left(\frac{10}{n} \cdot \frac{4}{4} + \frac{5}{4n}\right) = 2\left(\frac{40}{4n} + \frac{5}{4n}\right)$$

Combine the numerators for each expression.

$$2\left(\frac{40}{4n} + \frac{5}{4n}\right) = 2\left(\frac{40+5}{4n}\right) = 2\left(\frac{45}{4n}\right) = \frac{90}{4n}$$

Simplify the fraction.

$$\frac{90}{4n} = \frac{45}{2n}$$

The perimeter of the rectangle is $\frac{45}{2n}$.

53. A: The equation $3x + 1.50y = 300$ is written in standard form, we need to convert it into slope-intercept form, $y = mx + b$. First, move the x-term to the right side of the equation by subtracting $3x$ from both sides of the equation.

$$3x + 1.50y = 300$$
$$3x - 3x + 1.50y = 300 - 3x$$
$$1.50y = 300 - 3x$$

Next, isolate the variable y in the term $1.50y$ by dividing both sides of the equation by 1.50.

$$\frac{1.50y}{1.50} = \frac{300 - 3x}{1.50}$$
$$y = 200 - 2x$$

Finally, switch the order of the two terms on the right side of the equation. The slope-intercept form of the linear equation given is $y = -2x + 200$.

54. A: We will use long division to divide.

$$a - 2 \overline{)\, 2a^2 + 3a + 1}$$

Set up as a long division problem.

$$\begin{array}{r} 2a \\ a - 2 \overline{)\, 2a^2 + 3a + 1} \end{array}$$

Divide $2a^2$ by a.

$$\begin{array}{r} 2a \\ a - 2 \overline{)\, 2a^2 + 3a + 1} \\ -(2a^2 - 4a) \end{array}$$

Multiply $a - 2$ by $2a$ and write beneath $2a^2 + 3a$.

$$\begin{array}{r} 2a \\ a - 2 \overline{)\, 2a^2 + 3a + 1} \\ \underline{-(2a^2 - 4a)} \\ 7a \end{array}$$

Subtract the result from above to get a remainder of $7a$.

$$\begin{array}{r} 2a + 7 \\ a - 2 \overline{)\, 2a^2 + 3a + 1} \\ \underline{-(2a^2 - 4a)} \\ 7a + 1 \end{array}$$

Bring down the 1 and divide a into $7a$.

$$\begin{array}{r} 2a + 7 \\ a - 2 \overline{)\, 2a^2 + 3a + 1} \\ \underline{-(2a^2 - 4a)} \\ 7a + 1 \\ \underline{-\ (7a - 14)} \end{array}$$

Multiply $a - 2$ by 7 and write underneath $7a + 1$.

$$\begin{array}{r} 2a + 7 \\ a - 2 \overline{)\, 2a^2 + 3a + 1} \\ \underline{-(2a^2 - 4a)} \\ 7a + 1 \\ \underline{-\ (7a - 14)} \\ 15 \end{array}$$

Subtract the result from above to get a remainder of 15.

Dividing $2a^2 - 3a + 1$ by $a - 2$ yields a quotient of $2a + 7$ and a remainder of 15 which can be written as $2a + 7 + \frac{15}{a-2}$.

55. B: To figure out how much money Mia plans to deposit, use the formula $a_n = a_1 r^{(n-1)}$. First, identify the values of the variables used in the equation. August will be Mia's eighth deposit, so we are finding the eighth term in the geometric sequence. Since n is the number in the sequence that needs to be identified, replace n with 8 in the formula. The variable a_1 is the first term in the sequence. Mia's first deposit was in January when she opened her bank account with $10. Therefore, the first term in the sequence is 10. Each month, Mia plans to double the amount of money deposited. Each term gets doubled, or multiplied by 2, resulting in the subsequent term in the sequence. Based on this information, the common ratio r in this sequence is 2. Therefore, replace r with 2 in the formula.

$$a_n = a_1 r^{(n-1)}$$
$$a_8 = (10)(2)^{(8-1)}$$

Simplify using the order of operations.

$$a_8 = (10)(2)^{(7)}$$
$$a_8 = (10)(128)$$
$$a_8 = 1{,}280$$

Mia plans to deposit $1,280 into her bank account in August.

56. D: Start by finding the slope, m. To do so, use the slope formula $m = \frac{y_2 - y_1}{x_2 - x_1}$. Substitute the x- and y-values from the two coordinates given.

$$m = \frac{3 - (-2)}{-1 - 4} = \frac{5}{-5} = -1$$

Now that we know the slope is –1, we can write a point-slope equation using the slope and one of the points given. The point-slope equation of a line is $y - y_1 = m(x - x_1)$, where x_1 and y_1 are the coordinates of one point on the line, and m is the slope. Since we know the slope is –1, replace m with –1 in the equation. Next, substitute the x- and y-coordinates from one of the points into the equation. In this explanation, we will use point $(-1,3)$. The point-slope form of the equation is $y - 3 = -1(x - (-1))$.

Rearrange the equation to write it in standard form.

$$y - 3 = -1(x + 1)$$
$$y - 3 + 3 = -1x - 1 + 3$$
$$y = -1x + 2$$
$$x + y = 2$$

57. D: The table shows a linear relationship between t and M. To find an equation of this line, we choose two convenient points from the table, say $(1,73)$ and $(2,133)$, and use them to find the slope of the line: $m = \frac{133-73}{2-1} = \frac{60}{1} = 60$. Then we substitute this slope and either point into point-slope form, which looks like $M - M_1 = m(t - t_1)$ using these variables, and solve for M.

$$M - 73 = 60(t - 1)$$
$$M - 73 = 60t - 60$$
$$M = 60t + 13$$

Since each M-value on this line is the largest that makes the inequality true for the corresponding t-value, the rest are smaller. So, the inequality we seek is $M \leq 60t + 13$. In other words, if we must slow down due to traffic, the number of miles traveled will be less than the miles traveled if there was no traffic. The answer choices are in standard form, so rearrange this inequality to get $60t - M \geq -13$.

58. A: To find the x-intercepts for the graph of a quadratic function, we set $y = 0$ and solve for x. Setting $y = 0$ for our function, we get the quadratic equation, $0 = x^2 + 2x - 8$. The roots for the quadratic equation are also the x-intercepts of the corresponding graph.

While the discriminant for the quadratic formula will not tell you what the roots are, it can be used to determine how many roots exist. The discriminant for a quadratic equation in the standard form

$0 = ax^2 + bx + c$ is $b^2 - 4ac$, where a, b, and c are real numbers, and $a \neq 0$. Three conditions exist to determine how many roots a quadratic equation has using its discriminant.

$$b^2 - 4ac > 0 \quad\quad \text{There are two distinct real roots.}$$

$$b^2 - 4ac = 0 \quad\quad \text{There is one real root (often referred to as a double root).}$$

$$b^2 - 4ac < 0 \quad\quad \text{No real roots exist.}$$

For our quadratic equation $0 = x^2 + 2x - 8$, $a = 1$, $b = 2$, and $c = -8$. Find the discriminant.

$$b^2 - 4ac = (2)^2 - 4(1)(-8) = 4 + 32 = 36$$

Since the discriminant is positive in value, the quadratic equation has two distinct real roots, so the graph of the quadratic function $y = x^2 + 2x - 8$ crosses the x-axis two times.

59. C: The y-coordinate of the vertex for the graph of a quadratic function is either the minimum value or maximum value for the function. To find the vertex of the quadratic function, use the formula $x = -\frac{b}{2a}$ to find the x-coordinate, then plug this value into the original equation to find the y-coordinate.

$$x = \frac{-(-2)}{2(1)} = \frac{2}{2} = 1$$

Now, plug this in for x in the original equation and solve for y.

$$y = (1)^2 - 2(1) - 3 = 1 - 2 - 3 = -4$$

Therefore, the vertex of the parabola is at $(1, -4)$. Since the value of a is positive, the parabola opens up. This means the range of the function is $y \geq -4$.

60. D: A quadratic binomial in the form $a^2 - b^2$ is called a difference of two squares, where a^2 and b^2 are perfect squares, and factors to $(a - b)(a + b)$.

The two terms in our binomial are perfect squares and can be written as $(4s)^2 + (11)^2$, but it is not a difference of two perfect squares. It is a sum that cannot be factored.

61. A: Two lines are perpendicular if their slopes are opposite reciprocals of one another. To identify an equation of a line perpendicular to $x + 3y = 6$, start by identifying the slope, m, of the given line. To do so, write the equation in slope–intercept form, $y = mx + b$.

First, move the x-term to the right side of the equation by subtracting x from both sides of the equation.

$$3y = 6 - x$$

Next, divide both sides of the equation by 3.

$$\frac{3y}{3} = \frac{6 - x}{3}$$
$$y = 2 - \frac{1}{3}x$$

Switch the order of the two terms on the right side of the equation. The slope–intercept form of the linear equation given is $y = -\frac{1}{3}x + 2$.

Now that the equation is written in slope–intercept form, identify the slope. Since the value of m in this equation is $-\frac{1}{3}$, the slope of a perpendicular line must be the opposite reciprocal of $-\frac{1}{3}$, which is 3.

Write a new equation of a line in slope–intercept form, $y = mx + b$, and replace m with 3 in the equation.

$$y = 3x + b$$

Next, substitute the point $(9, -1)$ into the equation $y = 3x + b$.

$$(-1) = 3(9) + b$$

Simplify the equation by multiplying 3 and 9.

$$-1 = 27 + b$$

From here, solve for b. Subtract 27 from both sides of the equation.

$$-28 = b$$

Now that we know the slope is 3 and the y-intercept is –28, write the equation for this line in slope–intercept form, $y = mx + b$.

$$y = 3x - 28$$

The equation of a line that is perpendicular to the graph of $x + 3y = 6$ and contains point $(9, -1)$ is $y = 3x - 28$.

62. D: Decompose $11x$ to the sum of two terms so that one has a common factor with $3x^2$ and the other has a common factor with 10.

$$3x^2 + 6x + 5x + 10$$

Now, factor the greatest common factor (GCF) from the first two terms and the second two terms.

$$3x(x + 2) + 5(x + 2)$$

Finally, factor the GCF of $(x + 2)$ from the two groups.

$$(x + 2)(3x + 5)$$

63. B: $\frac{x}{2} > 3$ is an inequality that can be solved in one step. Treat the inequality as an equation, and isolate the variable using inverse operations. Multiply both sides by 2.

$$2\left(\frac{x}{2}\right) > 3(2)$$

$$x > 6$$

64. D: The equation of a line perpendicular to the x-axis runs vertically, passing through the x-axis. The x-coordinate of each point on this line is constant, represented by the variable k. Therefore, an equation of a line perpendicular to the x-axis is $x = k$. Based on the answer choices, k is equal to 8. Therefore, $x = 8$.

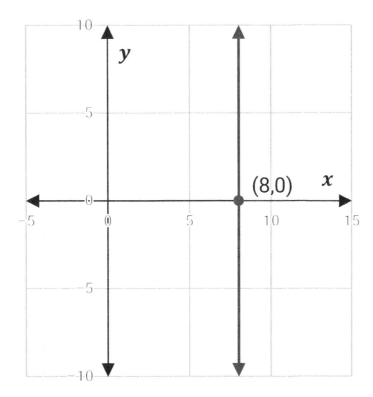

A vertical line always has an undefined slope. Recall that slope is rise over run, which is the change in y over the change in x. A vertical line will always have a change in x equal to 0. Since we can't divide by 0, its slope is undefined. Therefore, the slope of the line $x = 8$ is undefined.

65. C: The area, A, of a rectangle is $A = l \times w$, where l is the length of the rectangle, and w is the width of the rectangle.

Substituting the given values for the length, width, and area, we have the following equation.

$$63 = (x + 4)(x + 2)$$
$$63 = x^2 + 2x + 4x + 8$$
$$63 = x^2 + 6x + 8$$

This equation can be solved by factoring. First set the equation equal to 0 by subtracting 63 from both sides.

$$0 = x^2 + 6x - 55$$

Factor the quadratic on the right-hand side, then solve for x.

$$0 = x^2 + 6x - 55$$
$$0 = (x+11)(x-5)$$
$$x + 11 = 0 \text{ and } x - 5 = 0$$
$$x = -11 \text{ and } x = 5$$

While both answers satisfy the quadratic equation, only 5 satisfies the x-value for the possible dimensions of the rectangle since $x = 5$ yields positive values for the length and width of the rectangle, but $x = -11$ yields a length of –7 feet and a width of –9 feet, which is not possible.

66. B: To solve, start by finding the value of $f(8)$. Substitute 8 into the function $f(x) = x^2 - 3x$ by replacing the variable x with 8 and simplifying the expression.

$$f(x) = x^2 - 3x$$
$$f(8) = (8)^2 - 3(8)$$
$$f(8) = 64 - 3(8)$$
$$f(8) = 64 - 24$$
$$f(8) = 40$$

Now, find the value of $f(2)$. Substitute 2 into the function $f(x) = x^2 - 3x$ by replacing the variable x with 2, then simplify the expression.

$$f(x) = x^2 - 3x$$
$$f(2) = (2)^2 - 3(2)$$
$$f(2) = 4 - 3(2)$$
$$f(2) = 4 - 6$$
$$f(2) = -2$$

Now that we know the values of $f(8)$ and $f(2)$ in relation to the function $f(x) = x^2 - 3x$, we can find the value of $\frac{f(8)}{f(2)}$.

$$\frac{f(8)}{f(2)} = \frac{40}{-2} = -20$$

67. D: To find the answer, solve for a in the equation by isolating it on one side of the equation.

$$7a - 21 = 28$$

Start by adding 21 to both sides of the equation.

$$7a - 21 + 21 = 28 + 21$$
$$7a = 49$$

Next, divide both sides of the equation by 7.

$$\frac{7a}{7} = \frac{49}{7}$$
$$a = 7$$

68. C: The equation that represents the cost of Apartment A is $1,200m + 300$, and the equation that represents Apartment B is $1,050m + 600$. To find how many months before the cost is the same, set the equations equal to one another and solve for m.

$$1,200m + 300 = 1,050m + 600$$

First, subtract $1,050m$ from both sides of the equation.

$$1,200m - 1,050m + 300 = 1,050m - 1,050m + 600$$
$$150m + 300 = 600$$

Then, subtract 300 from both sides of the equation.

$$150m + 300 - 300 = 600 - 300$$
$$150m = 300$$

Finally, divide both sides of the equation by 150.

$$\frac{150m}{150} = \frac{300}{150}$$
$$m = 2$$

It will take 2 months for the costs of both apartments to be the same.

69. C: Let the variable b represent the price of each bag of food. Sue only has \$80 to spend, so the total cost of her dog food purchase needs to be less than or equal to 80. The inequality $5b \leq 80$ can be used to determine the maximum price per bag. To solve for the price per bag, divide both sides by 5.

$$b \leq 16$$

This means that Sue can afford 5 bags of dog food as long as each bag is less than or equal to \$16.

70. D: To find the value of y for the linear function when the value of x is 8, substitute 8 in for x and evaluate to find the value of y.

$$y = 5(8) + 3$$
$$y = 40 + 3$$
$$y = 43$$

71. D: A quadratic function in standard form is $y = ax^2 + bx + c$, where a, b, and c are real numbers and $a \neq 0$.

$y = 3(x-2)(x-2) + 1$	Square the binomial.
$y = 3(x^2 - 4x + 4) + 1$	Distribute using the FOIL method.
$y = 3x^2 - 12x + 12 + 1$	Distribute the coefficient of 3.
$y = 3x^2 - 12x + 13$	Combine like terms.

72. A: The equation of a parabola in vertex form is $y = a(x - h)^2 + k$, where (h, k) is the vertex of the parabola, and a is a real number that cannot be 0.

Since the projectile follows a parabolic path, its maximum height represents the y-coordinate of the vertex of the parabola, and the time at which this occurs represents the x-coordinate of the vertex. Substituting the vertex $(3,126)$ into the vertex form, we get the following equation.

$$h = a(t-3)^2 + 126$$

Also, since the projectile is launched from the ground, its height is 0 at time $t = 0$, so the point $(0,0)$ is another point on the graph of the parabola. Substitute $(0,0)$ in for x and y respectively to find the value for a.

$$0 = a(0-3)^2 + 126$$
$$0 = a(-3)^2 + 126$$
$$0 = 9a + 126$$
$$-126 = 9a$$
$$\frac{-126}{9} = a$$
$$-14 = a$$

In vertex form, the equation for the path of the projectile is $h = -14(t-3)^2 + 126$. Note the graph of the parabola opens downward, verifying the value for a is less than 0.

73. A: Start by creating a table that shows the relationship between hours walked (h) and distance (d).

Hours walked (h)	Distance in miles (d)
0	0
1	3
2	6
3	9

As you can see, the table represents the equation $d = 3h$. The independent variable is hours and the dependent variable is the distance walked in miles. From here, we can use the data in the table to graph points on the coordinate plane. The x-axis represents the independent variable, which is hours walked, and the y-axis represents the dependent variable, which is the distance walked in miles.

Rewrite the data from the table as points on the coordinate plane, $(x\ y)$: $(0,0), (1,3), (2,6), (3,9)$. The only graph with all four points is choice A.

74. C: Solve the equation for y by isolating the variable y on one side of the equation.

$$-x + 8y = -14$$

First, move x from the left side of the equation to the right by adding x to both sides of the equation.

$$-x + x + 8y = -14 + x$$
$$8y = -14 + x$$

Next, divide both sides of the equation by 8.

$$\frac{8y}{8} = \frac{-14+x}{8}$$
$$y = \frac{-14+x}{8}$$

75. C: First write the statement as an expression.

$$3n^2(4n^3+1)$$

Now, multiply the left side of the equation using the distributive property of multiplication.

$$3n^2(4n^3+1)$$

$$3n^2 \cdot 4n^3 + 3n^2 \cdot 1$$

$$12n^{2+3} + 3n^2$$

$$12n^5 + 3n^2$$

So, the product in terms of n is $12n^5 + 3n^2$.

Practice Tests #2 and #3

To take these additional practice tests, visit the bonus page for this product:
mometrix.com/bonus948/algebra1wb

Quiz Answer Key

For worked out answer explanations, please visit the bonus page for this product at **mometrix.com/bonus948/algebra1wb** or scan the QR code.

CHAPTER 1: SOLVING EQUATIONS

	1.1	1.2	1.3	1.4	1.5	1.6	1.7	1.8	1.9	1.10
1	C	C	D	A	C	B	D	A	B	C
2	A	D	B	A	B	B	B	B	D	C
3	A	A	C	B	A	D	C	C	D	C
4	B	A	D	A	A	A	D	A	B	D
5	D	B	A	C	B	C	C	C	C	B
6	A	A	A	B	A	D	A	D	D	A
7	D	D	B	A	B	D	A	B	B	C
8	B	C	B	B	D	A	C	C	A	D
9	B	D	C	B	B	C	D	C	D	A
10	A	D	A	B	D	D	A	B	A	C
11	D	B	B	A	A	A	C	C	B	D
12	B	A	A	A	D	A	B	B	D	B
13	A	B	D	B	C	C	B	B	C	A
14	B	C	D	B	B	B	A	C	D	B
15	D	A	D	C	C	A	D	B	A	A

CHAPTER 2: FUNCTIONS AND SEQUENCES

	2.1	2.2	2.3	2.4	2.5	2.6
1	A	C	A	B	C	D
2	D	B	B	A	B	A
3	B	A	B	C	B	A
4	A	D	A	C	C	D
5	A	A	C	B	A	A
6	B	B	C	B	C	A
7	B	A	B	C	A	B
8	C	C	A	B	C	B
9	A	B	B	D	D	A
10	D	D	D	C	B	D
11	D	B	C	A	B	C
12	B	B	C	A	C	A
13	C	A	A	B	C	C
14	B	C	D	C	D	A
15	C	C	D	A	D	B

CHAPTER 3: LINEAR EQUATIONS

	3.1	3.2	3.3	3.4	3.5	3.6	3.7	3.8	3.9	3.10	3.11	3.12
1	D	C	A	D	D	B	A	A	B	C	B	D
2	A	B	A	D	B	B	A	C	B	C	C	A
3	D	B	B	A	D	B	D	D	C	B	D	D
4	A	A	A	C	A	C	C	C	D	C	A	A
5	C	D	D	A	C	A	B	D	C	D	B	B
6	C	A	C	B	C	B	C	A	A	A	B	B
7	B	C	C	D	B	B	C	D	A	D	C	C
8	A	D	D	D	C	C	A	A	D	B	A	A
9	C	C	B	D	D	C	D	B	C	B	D	B
10	B	B	A	C	A	D	B	B	D	D	A	D
11	D	A	C	A	D	A	B	C	B	C	A	B
12	D	A	D	D	C	A	C	B	B	C	C	C
13	B	D	B	B	C	D	A	A	A	A	B	C
14	C	C	B	D	B	B	D	D	C	B	D	A
15	A	B	A	D	D	C	C	C	C	B	D	D

CHAPTER 4: INEQUALITIES

	4.1	4.2	4.3	4.4	4.5	4.6	4.7
1	C	A	A	C	A	A	A
2	D	A	B	A	B	A	C
3	C	D	A	D	D	D	D
4	B	D	D	A	B	A	A
5	B	B	C	B	A	C	B
6	A	C	C	D	C	B	C
7	C	B	D	C	C	B	A
8	D	D	A	B	D	A	C
9	B	C	B	B	B	D	D
10	D	C	B	C	D	C	A
11	A	B	D	A	B	B	C
12	B	D	B	D	B	D	B
13	B	B	D	B	C	C	D
14	C	A	A	C	D	A	C
15	C	B	C	A	D	C	B

Quiz Answer Key

For worked out answer explanations, please visit the bonus page for this product at **mometrix.com/bonus948/algebra1wb** or scan the QR code.

CHAPTER 1: SOLVING EQUATIONS

	1.1	1.2	1.3	1.4	1.5	1.6	1.7	1.8	1.9	1.10
1	C	C	D	A	C	B	D	A	B	C
2	A	D	B	A	B	B	B	B	D	C
3	A	A	C	B	A	D	C	C	D	C
4	B	A	D	A	A	A	D	A	B	D
5	D	B	A	C	B	C	C	C	C	B
6	A	A	A	B	A	D	A	D	D	A
7	D	D	B	A	B	D	A	B	B	C
8	B	C	B	B	D	A	C	C	A	D
9	B	D	C	B	B	C	D	C	D	A
10	A	D	A	B	D	D	A	B	A	C
11	D	B	B	A	A	A	C	C	B	D
12	B	A	A	A	D	A	B	B	D	B
13	A	B	D	B	C	C	B	B	C	A
14	B	C	D	B	B	B	A	C	D	B
15	D	A	D	C	C	A	D	B	A	A

CHAPTER 2: FUNCTIONS AND SEQUENCES

	2.1	2.2	2.3	2.4	2.5	2.6
1	A	C	A	B	C	D
2	D	B	B	A	B	A
3	B	A	B	C	B	A
4	A	D	A	C	C	D
5	A	A	C	B	A	A
6	B	B	C	B	C	A
7	B	A	B	C	A	B
8	C	C	A	B	C	B
9	A	B	B	D	D	A
10	D	D	D	C	B	D
11	D	B	C	A	B	C
12	B	B	C	A	C	A
13	C	A	A	B	C	C
14	B	C	D	C	D	A
15	C	C	D	A	D	B

CHAPTER 3: LINEAR EQUATIONS

	3.1	3.2	3.3	3.4	3.5	3.6	3.7	3.8	3.9	3.10	3.11	3.12
1	D	C	A	D	D	B	A	A	B	C	B	D
2	A	B	A	D	B	B	A	C	B	C	C	A
3	D	B	B	A	D	B	D	D	C	B	D	D
4	A	A	A	C	A	C	C	C	D	C	A	A
5	C	D	D	A	C	A	B	D	C	D	B	B
6	C	A	C	B	C	B	C	A	A	A	B	B
7	B	C	C	D	B	B	C	D	A	D	C	C
8	A	D	D	D	C	C	A	A	D	B	A	A
9	C	C	B	D	D	C	D	B	C	B	D	B
10	B	B	A	C	A	D	B	B	D	D	A	D
11	D	A	C	A	D	A	B	C	B	C	A	B
12	D	A	D	D	C	A	C	B	B	C	C	C
13	B	D	B	B	C	D	A	A	A	A	B	C
14	C	C	B	D	B	B	D	D	C	B	D	A
15	A	B	A	D	D	C	C	C	C	B	D	D

CHAPTER 4: INEQUALITIES

	4.1	4.2	4.3	4.4	4.5	4.6	4.7
1	C	A	A	C	A	A	A
2	D	A	B	A	B	A	C
3	C	D	A	D	D	D	D
4	B	D	D	A	B	A	A
5	B	B	C	B	A	C	B
6	A	C	C	D	C	B	C
7	C	B	D	C	C	B	A
8	D	D	A	B	D	A	C
9	B	C	B	B	B	D	D
10	D	C	B	C	D	C	A
11	A	B	D	A	B	B	C
12	B	D	B	D	B	D	B
13	B	B	D	B	C	C	D
14	C	A	A	C	D	A	C
15	C	B	C	A	D	C	B

CHAPTER 5: SYSTEMS OF EQUATIONS

	5.1	5.2	5.3	5.4	5.5
1	A	D	A	A	B
2	C	B	C	C	B
3	C	C	B	D	D
4	B	D	D	B	C
5	C	C	C	D	B
6	B	A	A	A	A
7	D	B	D	A	D
8	A	C	D	C	B
9	B	C	B	B	C
10	C	B	A	D	A
11	A	D	C	C	D
12	D	D	C	D	A
13	B	A	B	B	C
14	C	B	A	C	B
15	B	C	B	A	A

CHAPTER 6: POLYNOMIALS

	6.1	6.2	6.3	6.4	6.5	6.6	6.7	6.8	6.9	6.10	6.11	6.12
1	C	B	C	D	B	B	C	B	C	C	D	B
2	A	C	D	B	C	D	C	A	A	B	A	B
3	D	A	C	B	D	A	B	B	D	D	D	A
4	B	C	D	A	D	B	A	A	B	B	D	D
5	D	D	C	B	B	C	C	A	A	C	A	C
6	B	B	B	D	C	A	D	A	D	D	C	C
7	B	A	B	D	B	D	A	D	D	C	A	B
8	B	D	D	D	C	A	D	C	C	A	B	C
9	B	C	B	A	D	C	B	B	B	C	D	B
10	C	A	C	B	D	B	C	A	A	B	C	A
11	D	B	D	C	B	C	C	B	B	B	B	B
12	A	D	B	A	A	D	C	D	D	D	A	C
13	B	C	B	C	C	A	D	B	D	C	C	C
14	A	B	D	D	C	C	D	A	C	D	B	D
15	D	A	B	B	D	D	B	B	C	C	C	A

CHAPTER 7: ALGEBRAIC FRACTIONS

	7.1	7.2	7.3	7.4	7.5	7.6	7.7
1	D	C	A	D	B	B	D
2	A	B	C	A	B	D	B
3	B	D	D	A	C	B	A
4	C	A	C	C	A	A	C
5	D	B	C	A	D	C	D
6	D	A	B	B	C	C	C
7	A	B	A	C	C	D	C
8	D	D	A	C	A	B	A
9	C	A	A	D	D	A	B
10	D	D	B	B	D	A	D
11	A	B	A	C	B	A	B
12	B	D	C	D	C	C	B
13	A	C	A	A	D	B	C
14	B	C	C	A	A	D	D
15	C	B	C	C	D	D	C

CHAPTER 8: QUADRATIC FUNCTIONS

	8.1	8.2	8.3	8.4	8.5	8.6	8.7	8.8	8.9	8.10	8.11
1	A	B	B	D	A	B	D	B	C	B	B
2	A	D	C	A	B	A	B	C	B	C	D
3	C	A	B	D	C	D	B	C	D	D	B
4	C	C	D	A	C	D	A	D	B	C	A
5	C	B	B	C	D	B	C	A	D	A	A
6	B	B	A	B	C	B	A	D	A	D	C
7	B	D	A	B	B	C	B	D	A	B	B
8	D	C	C	A	D	A	D	A	A	D	A
9	A	D	A	D	A	C	A	B	C	D	B
10	C	A	D	C	A	D	A	D	B	C	D
11	D	A	D	C	C	B	D	C	C	C	C
12	D	B	B	C	D	D	C	C	C	A	A
13	B	B	C	B	C	A	C	B	A	B	D
14	A	D	C	A	B	A	A	B	C	C	D
15	C	A	D	D	B	C	B	D	C	D	A

Chapter 9: Exponential Functions

	9.1	9.2	9.3
1	C	B	B
2	A	A	B
3	D	D	C
4	D	C	C
5	B	D	B
6	C	A	A
7	D	B	A
8	B	C	D
9	B	C	B
10	D	D	D
11	D	A	D
12	C	D	C
13	C	D	A
14	A	B	D
15	A	C	C

Chapter 10: Radical Expressions

	10.1	10.2	10.3	10.4	10.5	10.6	10.7
1	C	C	C	B	D	D	B
2	D	D	D	A	A	B	C
3	B	A	D	B	C	B	C
4	C	D	A	A	A	A	B
5	C	A	B	C	C	C	D
6	B	D	B	D	D	A	A
7	D	A	C	B	C	A	A
8	A	D	B	A	D	B	A
9	B	D	C	A	D	D	D
10	B	A	A	B	C	A	C
11	A	D	A	D	B	C	B
12	A	B	B	C	D	C	B
13	A	B	B	A	C	B	A
14	B	D	C	B	A	B	D
15	D	A	A	B	A	B	C

For worked out answer explanations, please visit the bonus page for this product at **mometrix.com/bonus948/algebra1wb** or scan the QR code.

How to Overcome Test Anxiety

Just the thought of taking a test is enough to make most people a little nervous. A test is an important event that can have a long-term impact on your future, so it's important to take it seriously and it's natural to feel anxious about performing well. But just because anxiety is normal, that doesn't mean that it's helpful in test taking, or that you should simply accept it as part of your life. Anxiety can have a variety of effects. These effects can be mild, like making you feel slightly nervous, or severe, like blocking your ability to focus or remember even a simple detail.

If you experience test anxiety—whether severe or mild—it's important to know how to beat it. To discover this, first you need to understand what causes test anxiety.

Causes of Test Anxiety

While we often think of anxiety as an uncontrollable emotional state, it can actually be caused by simple, practical things. One of the most common causes of test anxiety is that a person does not feel adequately prepared for their test. This feeling can be the result of many different issues such as poor study habits or lack of organization, but the most common culprit is time management. Starting to study too late, failing to organize your study time to cover all of the material, or being distracted while you study will mean that you're not well prepared for the test. This may lead to cramming the night before, which will cause you to be physically and mentally exhausted for the test. Poor time management also contributes to feelings of stress, fear, and hopelessness as you realize you are not well prepared but don't know what to do about it.

Other times, test anxiety is not related to your preparation for the test but comes from unresolved fear. This may be a past failure on a test, or poor performance on tests in general. It may come from comparing yourself to others who seem to be performing better or from the stress of living up to expectations. Anxiety may be driven by fears of the future—how failure on this test would affect your educational and career goals. These fears are often completely irrational, but they can still negatively impact your test performance.

Elements of Test Anxiety

As mentioned earlier, test anxiety is considered to be an emotional state, but it has physical and mental components as well. Sometimes you may not even realize that you are suffering from test anxiety until you notice the physical symptoms. These can include trembling hands, rapid heartbeat, sweating, nausea, and tense muscles. Extreme anxiety may lead to fainting or vomiting. Obviously, any of these symptoms can have a negative impact on testing. It is important to recognize them as soon as they begin to occur so that you can address the problem before it damages your performance.

The mental components of test anxiety include trouble focusing and inability to remember learned information. During a test, your mind is on high alert, which can help you recall information and stay focused for an extended period of time. However, anxiety interferes with your mind's natural processes, causing you to blank out, even on the questions you know well. The strain of testing during anxiety makes it difficult to stay focused, especially on a test that may take several hours. Extreme anxiety can take a huge mental toll, making it difficult not only to recall test information but even to understand the test questions or pull your thoughts together.

Effects of Test Anxiety

Test anxiety is like a disease—if left untreated, it will get progressively worse. Anxiety leads to poor performance, and this reinforces the feelings of fear and failure, which in turn lead to poor performances on subsequent tests. It can grow from a mild nervousness to a crippling condition. If allowed to progress, test anxiety can have a big impact on your schooling, and consequently on your future.

Test anxiety can spread to other parts of your life. Anxiety on tests can become anxiety in any stressful situation, and blanking on a test can turn into panicking in a job situation. But fortunately, you don't have to let anxiety rule your testing and determine your grades. There are a number of relatively simple steps you can take to move past anxiety and function normally on a test and in the rest of life.

Physical Steps for Beating Test Anxiety

While test anxiety is a serious problem, the good news is that it can be overcome. It doesn't have to control your ability to think and remember information. While it may take time, you can begin taking steps today to beat anxiety.

Just as your first hint that you may be struggling with anxiety comes from the physical symptoms, the first step to treating it is also physical. Rest is crucial for having a clear, strong mind. If you are tired, it is much easier to give in to anxiety. But if you establish good sleep habits, your body and mind will be ready to perform optimally, without the strain of exhaustion. Additionally, sleeping well helps you to retain information better, so you're more likely to recall the answers when you see the test questions.

Getting good sleep means more than going to bed on time. It's important to allow your brain time to relax. Take study breaks from time to time so it doesn't get overworked, and don't study right before bed. Take time to rest your mind before trying to rest your body, or you may find it difficult to fall asleep.

Along with sleep, other aspects of physical health are important in preparing for a test. Good nutrition is vital for good brain function. Sugary foods and drinks may give a burst of energy but this burst is followed by a crash, both physically and emotionally. Instead, fuel your body with protein and vitamin-rich foods.

Also, drink plenty of water. Dehydration can lead to headaches and exhaustion, especially if your brain is already under stress from the rigors of the test. Particularly if your test is a long one, drink water during the breaks. And if possible, take an energy-boosting snack to eat between sections.

Along with sleep and diet, a third important part of physical health is exercise. Maintaining a steady workout schedule is helpful, but even taking 5-minute study breaks to walk can help get your blood pumping faster and clear your head. Exercise also releases endorphins, which contribute to a positive feeling and can help combat test anxiety.

When you nurture your physical health, you are also contributing to your mental health. If your body is healthy, your mind is much more likely to be healthy as well. So take time to rest, nourish your body with healthy food and water, and get moving as much as possible. Taking these physical steps will make you stronger and more able to take the mental steps necessary to overcome test anxiety.

Mental Steps for Beating Test Anxiety

Working on the mental side of test anxiety can be more challenging, but as with the physical side, there are clear steps you can take to overcome it. As mentioned earlier, test anxiety often stems from lack of preparation, so the obvious solution is to prepare for the test. Effective studying may be the most important weapon you have for beating test anxiety, but you can and should employ several other mental tools to combat fear.

First, boost your confidence by reminding yourself of past success—tests or projects that you aced. If you're putting as much effort into preparing for this test as you did for those, there's no reason you should expect to fail here. Work hard to prepare; then trust your preparation.

Second, surround yourself with encouraging people. It can be helpful to find a study group, but be sure that the people you're around will encourage a positive attitude. If you spend time with others who are anxious or cynical, this will only contribute to your own anxiety. Look for others who are motivated to study hard from a desire to succeed, not from a fear of failure.

Third, reward yourself. A test is physically and mentally tiring, even without anxiety, and it can be helpful to have something to look forward to. Plan an activity following the test, regardless of the outcome, such as going to a movie or getting ice cream.

When you are taking the test, if you find yourself beginning to feel anxious, remind yourself that you know the material. Visualize successfully completing the test. Then take a few deep, relaxing breaths and return to it. Work through the questions carefully but with confidence, knowing that you are capable of succeeding.

Developing a healthy mental approach to test taking will also aid in other areas of life. Test anxiety affects more than just the actual test—it can be damaging to your mental health and even contribute to depression. It's important to beat test anxiety before it becomes a problem for more than testing.

Study Strategy

Being prepared for the test is necessary to combat anxiety, but what does being prepared look like? You may study for hours on end and still not feel prepared. What you need is a strategy for test prep. The next few pages outline our recommended steps to help you plan out and conquer the challenge of preparation.

STEP 1: SCOPE OUT THE TEST

Learn everything you can about the format (multiple choice, essay, etc.) and what will be on the test. Gather any study materials, course outlines, or sample exams that may be available. Not only will this help you to prepare, but knowing what to expect can help to alleviate test anxiety.

STEP 2: MAP OUT THE MATERIAL

Look through the textbook or study guide and make note of how many chapters or sections it has. Then divide these over the time you have. For example, if a book has 15 chapters and you have five days to study, you need to cover three chapters each day. Even better, if you have the time, leave an extra day at the end for overall review after you have gone through the material in depth.

If time is limited, you may need to prioritize the material. Look through it and make note of which sections you think you already have a good grasp on, and which need review. While you are studying, skim quickly through the familiar sections and take more time on the challenging parts.

Write out your plan so you don't get lost as you go. Having a written plan also helps you feel more in control of the study, so anxiety is less likely to arise from feeling overwhelmed at the amount to cover.

STEP 3: GATHER YOUR TOOLS

Decide what study method works best for you. Do you prefer to highlight in the book as you study and then go back over the highlighted portions? Or do you type out notes of the important information? Or is it helpful to make flashcards that you can carry with you? Assemble the pens, index cards, highlighters, post-it notes, and any other materials you may need so you won't be distracted by getting up to find things while you study.

If you're having a hard time retaining the information or organizing your notes, experiment with different methods. For example, try color-coding by subject with colored pens, highlighters, or post-it notes. If you learn better by hearing, try recording yourself reading your notes so you can listen while in the car, working out, or simply sitting at your desk. Ask a friend to quiz you from your flashcards, or try teaching someone the material to solidify it in your mind.

STEP 4: CREATE YOUR ENVIRONMENT

It's important to avoid distractions while you study. This includes both the obvious distractions like visitors and the subtle distractions like an uncomfortable chair (or a too-comfortable couch that makes you want to fall asleep). Set up the best study environment possible: good lighting and a comfortable work area. If background music helps you focus, you may want to turn it on, but otherwise keep the room quiet. If you are using a computer to take notes, be sure you don't have any other windows open, especially applications like social media, games, or anything else that could distract you. Silence your phone and turn off notifications. Be sure to keep water close by so you stay hydrated while you study (but avoid unhealthy drinks and snacks).

Also, take into account the best time of day to study. Are you freshest first thing in the morning? Try to set aside some time then to work through the material. Is your mind clearer in the afternoon or evening? Schedule your study session then. Another method is to study at the same time of day that you will take the test, so that your brain gets used to working on the material at that time and will be ready to focus at test time.

STEP 5: STUDY!

Once you have done all the study preparation, it's time to settle into the actual studying. Sit down, take a few moments to settle your mind so you can focus, and begin to follow your study plan. Don't give in to distractions or let yourself procrastinate. This is your time to prepare so you'll be ready to fearlessly approach the test. Make the most of the time and stay focused.

Of course, you don't want to burn out. If you study too long you may find that you're not retaining the information very well. Take regular study breaks. For example, taking five minutes out of every hour to walk briskly, breathing deeply and swinging your arms, can help your mind stay fresh.

As you get to the end of each chapter or section, it's a good idea to do a quick review. Remind yourself of what you learned and work on any difficult parts. When you feel that you've mastered the material, move on to the next part. At the end of your study session, briefly skim through your notes again.

But while review is helpful, cramming last minute is NOT. If at all possible, work ahead so that you won't need to fit all your study into the last day. Cramming overloads your brain with more information than it can process and retain, and your tired mind may struggle to recall even previously learned information when it is overwhelmed with last-minute study. Also, the urgent nature of cramming and the stress placed on your brain contribute to anxiety. You'll be more likely to go to the test feeling unprepared and having trouble thinking clearly.

So don't cram, and don't stay up late before the test, even just to review your notes at a leisurely pace. Your brain needs rest more than it needs to go over the information again. In fact, plan to finish your studies by noon or early afternoon the day before the test. Give your brain the rest of the day to relax or focus on other things, and get a good night's sleep. Then you will be fresh for the test and better able to recall what you've studied.

STEP 6: TAKE A PRACTICE TEST

Many courses offer sample tests, either online or in the study materials. This is an excellent resource to check whether you have mastered the material, as well as to prepare for the test format and environment.

Check the test format ahead of time: the number of questions, the type (multiple choice, free response, etc.), and the time limit. Then create a plan for working through them. For example, if you have 30 minutes to take a 60-question test, your limit is 30 seconds per question. Spend less time on the questions you know well so that you can take more time on the difficult ones.

If you have time to take several practice tests, take the first one open book, with no time limit. Work through the questions at your own pace and make sure you fully understand them. Gradually work up to taking a test under test conditions: sit at a desk with all study materials put away and set a timer. Pace yourself to make sure you finish the test with time to spare and go back to check your answers if you have time.

After each test, check your answers. On the questions you missed, be sure you understand why you missed them. Did you misread the question (tests can use tricky wording)? Did you forget the information? Or was it something you hadn't learned? Go back and study any shaky areas that the practice tests reveal.

Taking these tests not only helps with your grade, but also aids in combating test anxiety. If you're already used to the test conditions, you're less likely to worry about it, and working through tests until you're scoring well gives you a confidence boost. Go through the practice tests until you feel comfortable, and then you can go into the test knowing that you're ready for it.

Test Tips

On test day, you should be confident, knowing that you've prepared well and are ready to answer the questions. But aside from preparation, there are several test day strategies you can employ to maximize your performance.

First, as stated before, get a good night's sleep the night before the test (and for several nights before that, if possible). Go into the test with a fresh, alert mind rather than staying up late to study.

Try not to change too much about your normal routine on the day of the test. It's important to eat a nutritious breakfast, but if you normally don't eat breakfast at all, consider eating just a protein bar. If you're a coffee drinker, go ahead and have your normal coffee. Just make sure you time it so that the caffeine doesn't wear off right in the middle of your test. Avoid sugary beverages, and drink enough water to stay hydrated but not so much that you need a restroom break 10 minutes into the test. If your test isn't first thing in the morning, consider going for a walk or doing a light workout before the test to get your blood flowing.

Allow yourself enough time to get ready, and leave for the test with plenty of time to spare so you won't have the anxiety of scrambling to arrive in time. Another reason to be early is to select a good seat. It's helpful to sit away from doors and windows, which can be distracting. Find a good seat, get out your supplies, and settle your mind before the test begins.

When the test begins, start by going over the instructions carefully, even if you already know what to expect. Make sure you avoid any careless mistakes by following the directions.

Then begin working through the questions, pacing yourself as you've practiced. If you're not sure on an answer, don't spend too much time on it, and don't let it shake your confidence. Either skip it and come back later, or eliminate as many wrong answers as possible and guess among the remaining ones. Don't dwell on these questions as you continue—put them out of your mind and focus on what lies ahead.

Be sure to read all of the answer choices, even if you're sure the first one is the right answer. Sometimes you'll find a better one if you keep reading. But don't second-guess yourself if you do immediately know the answer. Your gut instinct is usually right. Don't let test anxiety rob you of the information you know.

If you have time at the end of the test (and if the test format allows), go back and review your answers. Be cautious about changing any, since your first instinct tends to be correct, but make sure you didn't misread any of the questions or accidentally mark the wrong answer choice. Look over any you skipped and make an educated guess.

At the end, leave the test feeling confident. You've done your best, so don't waste time worrying about your performance or wishing you could change anything. Instead, celebrate the successful completion of this test. And finally, use this test to learn how to deal with anxiety even better next time.

Review Video: Test Anxiety
Visit mometrix.com/academy and enter code: 100340

Important Qualification

Not all anxiety is created equal. If your test anxiety is causing major issues in your life beyond the classroom or testing center, or if you are experiencing troubling physical symptoms related to your anxiety, it may be a sign of a serious physiological or psychological condition. If this sounds like your situation, we strongly encourage you to seek professional help.

Additional Bonus Material

Due to our efforts to try to keep this book to a manageable length, we've created a link that will give you access to all of your additional bonus material:

mometrix.com/bonus948/algebra1wb

Made in the USA
Middletown, DE
01 June 2025

76388072R00243